Technology Matters

Technology Matters

Questions to Live With

David E. Nye

The MIT Press
Cambridge, Massachusetts
London, England

First MIT Press paperback edition, 2007

© 2006 Massachusetts Institute of Technology

MIT Press books may be purchased at special quantity discounts for business or sales promotional use. For information, please email special_sales@mitpress.mit.edu.

Set in Stone Sans and Stone Serif by Graphic Composition, Inc.
Printed and bound in the United States of America.

Library of Congress Cataloging-in-Publication Data

Nye, David E., 1946–
Technology matters: questions to live with / David E. Nye
   p.   cm.
Includes bibliographical references and index.
ISBN-10: 0-262-14093-4 (hc: alk. paper)—0-262-64067-8 (pb: alk. paper)
ISBN-13: 978-0-262-14093-5 (hc: alk. paper)—978-0-262-64067-1
   (pb: alk. paper)
1. Technology—Philosophy. 2. Technology and civilization. I. Title.
T14.N88   2006   303.48'3—dc22   2005052114

10 9 8 7 6 5 4

Contents

Preface

Technology matters because it is inseparable from being human. Devices and machines are not things "out there" that invade life. We are intimate with them from birth, as were our ancestors for hundreds of generations. Like most children born in the twentieth century, I played with technological toys—miniature trucks, cars, stoves, airplanes, and railroads and full-size fake guns, swords, and telephones. With such toys I built castles, reshaped landscapes, put out imaginary fires, fought bloodless wars, and prepared imaginary food. Children learn to conceive technological solutions to problems, and in doing so they shape their own imaginations. Computer games add new dimensions to this process, but the fundamental point remains the same: By playing with technological toys, boys and girls imagine themselves into a creative relationship with the world. For a few people, playful imitation leads directly to a life's work as a fireman, an architect, a truck driver, a pilot, a soldier, a cook, a farmer, or a mechanic. These people are exceptions. Yet as adults many people retain their technological playfulness, expressing it in the acquisition of new appliances, gadgets, software, car accessories, and sports equipment. We live not merely in a technological world, but in a world that from our earliest years we imagine and construct through tools and machines.

Yet we seldom think systematically about the machines and systems that surround us. Do they shape us, or we them? How is our relationship to them changing? Are machines in the saddle, or will they soon be? Do they enrich us, or impoverish us? Are we using them more to destroy the natural world, or more to protect it? To undermine democracy, or to enhance it? To homogenize cultural differences, or increase them? To make the world more secure, or more dangerous? These are some of the questions that led me to write this book. I examine some of the more persuasive answers, but I can provide only provisional solutions. As with all large issues, technological questions resist final answers. When a young poet wrote to Rainer Maria Rilke that he had lost his faith in God, Rilke famously replied: ". . . be patient toward all that is unsolved in your heart and try to love the questions themselves like locked rooms and like books that are written in a very foreign tongue. . . . Live the questions now. Perhaps you will then gradually, without noticing it, live along some distant day into the answers." (*Letters to a Young Poet,* revised edition, Norton, 1954, p. 35)

I am asking readers of this book to recognize and wrestle with technological questions. The toys of childhood gave us easy solutions to imaginary problems. As adults, we see the complexity of technology, including its side effects and its unintended consequences as well as its benefits. Like it or not, we must live along until the day when we can understand technology better.

I devote a chapter each to the following questions, which I consider central and unavoidable:

- What is technology?
- Is technology inherently deterministic, or is it inflected or even shaped by culture?
- Is technology predictable?

- How do historians understand technology?
- Does using modern technologies break down cultural differences, or does it increase them?
- What is the relationship between technology and nature?
- Do new technologies destroy jobs, or do they create new opportunities?
- How should societies choose new technologies? Should "the market" decide?
- Do advanced technologies make life more secure, or do they expose humanity to escalating dangers?
- Does increasing use of technologies expand mental horizons, or does it encapsulate human beings in artifice?

This book aims to help people think about life in an intensely technological world. It addresses general questions that ordinary people ask but specialists often overlook. Scholars, understandably, prefer to avoid such large, general questions and instead to work on clearly defined, manageable subjects. Since it is hard to give definite answers to these big questions, it is safer not to attempt to do so. Yet these large and perhaps unanswerable questions are important. So this book is not necessarily for specialists; it is for anyone who has wondered about such things. My answers are provisional, until readers find better ones.

Acknowledgements

Every book builds upon an author's previous research, and compiling acknowledgements becomes more difficult with every new project. Several institutions made their libraries available to me, particularly Warwick University, Cambridge University, the British Library, the Massachusetts Institute of Technology, Notre Dame University, and the University of Southern Denmark. Most of the material in this book has not appeared before. However, chapter 3 first appeared in somewhat different form in *Technological Visions: The Hopes and Fears That Shape New Technologies,* edited by Marita Sturken, Douglas Thomas, and Sandra J. Ball-Rokeach (Temple University Press, 2004), and early versions of small portions of several other chapters appeared in my essay "Critics of Technology" in *A Companion to American Technology,* edited by Carroll Pursell (Blackwell, 2005).

For decades, many people have helped clarify my thinking about the subjects treated here. I cannot begin to name them all or to make use of all their good advice. I owe a general debt to the Society for the History of Technology for providing a stimulating climate of discussion and a wealth of excellent papers at its annual meetings. More specifically, I thank Leo Marx, Larry Cohen, Bob Friedel, Roe Smith, Cecilia Tichi, Miles Orvell, Carroll Pursell,

Roz Williams, Bob Gross, Marty Melosi, Bernie Carlson, and Tom Hughes. All of them, in many short but often intense conversations, shared with me their wide knowledge without entirely puncturing my illusions of competence. As with my earlier work, Helle Bertramsen Nye endured and encouraged my incessant talk about this project. Finally, I thank my father, Edwin P. Nye (1920–2004), a retired professor of mechanical engineering at Trinity College in Hartford, who passed away as this work neared completion. I never learned all that he had to teach about technology and society, but he imparted many insights that became part of both this and my earlier work. I am proud to dedicate this book to his memory.

Technology Matters

1 | Can We Define "Technology"?

One way to define "technology" is in terms of evolution. An animal may briefly use a natural object, such as a branch or a stone, for a purpose, but it was long thought that only human beings intentionally made objects, such as a rake or a hammer, for certain functions. Benjamin Franklin and many others thought that tool use separated humans from all other creatures. Recent fieldwork complicates the picture. Jane Goodall watched a chimpanzee in its own habitat. It found a twig of a certain size, peeled off its bark, looked for a termite hill, thrust in the peeled twig, pulled it out covered with termites, and ate them. This chimpanzee not only made a tool, it did so with forethought. In 2004, scientists announced discovery of the bones of a previously unknown species in an Indonesian cave. Standing only three feet high, this dwarf species lived and used tools as recently as 12,000 years ago.[1] Yet if Franklin's idea needs modification, it seems that only intelligent apes and human species are toolmakers, while the vast majority of animals are not. Birds construct nests. Beavers cut down trees and build dams. Ants and bees build complex communities that include a division of labor and food storage. But only a few species have made tools. Notable is a hand axe widely used by *Homo erectus* 1.6 million years ago.

Homo sapiens have used tools for at least 400,000 years, and seem to have done so from their first emergence. Technologies are not foreign to "human nature" but inseparable from it. Our ancestors evolved an opposition between thumb and fingers that made it easier to grasp and control objects than it is for other species. Indeed, prehensile hands may even have evolved simultaneously with the enlarging human cortex. Learning to use tools was a crucial step in the species' development, both because it increased adaptability and because it led to a more complex social life. Using tools, the relatively weak *Homo sapiens* were able to capture and domesticate animals, create and control fire, fashion artifacts, build shelters, and kill large animals. Deadly tools also facilitated murder and warfare. Tools emerged with the higher apes, and one might argue that humanity fashioned itself with tools.[2]

The central purpose of technologies has not been to provide necessities, such as food and shelter, for humans had achieved these goals very early in their existence. Rather, technologies have been used for social evolution. "Technology," José Ortega y Gasset argued, "is the production of superfluities—today as in the Paleolithic age. That is why animals are atechnical; they are content with the simple act of living."[3] Humans, in contrast, continually redefine their necessities to include more. Necessity is often not the mother of invention. In many cases, it surely has been just the opposite, and invention has been the mother of necessity. When humans possess a tool, they excel at finding new uses for it. The tool often exists before the problem to be solved. Latent in every tool are unforeseen transformations.

Defining technology as inseparable from human evolution suggests that tools and machines are far more than objects whose meaning is revealed simply by their purposes. As the great stone circle at Stonehenge reminds us, they are part of systems of meaning, and they express larger sequences of actions and ideas. Ultimately, the meaning of a tool is inseparable from the stories that surround it. Consider the similarity between what is involved in creating and using a tool and the sequence of a narrative. Even the chimpanzee picking up and peeling a twig to "fish" for termites requires the mental projection of a sequence, including an initial desire, several actions, and successful feeding. The sequence becomes more complex when more tools are involved, or when the same tool is used in several ways. Composing a narrative and using a tool are not identical processes, but they have affinities. Each requires the imagination of altered circumstances, and in each case beings must see themselves to be living in time. Making a tool immediately implies a succession of events in which one exercises some control over outcomes. Either to tell a story or to make a tool is to adopt an imaginary position outside immediate sensory experience. In each case, one imagines how present circumstances might be made different.

When faced with an inadvertently locked automobile with the keys inside, for example, one has a problem with several possible solutions—in effect, a story with several potential endings. One could call a locksmith, or one could use a rock to break one of the car's windows. Neither is as elegant a solution as passing a twisted coat hanger through a slightly open window and lifting the door handle from the inside. To improvise with tools or to tell stories requires the ability to imagine not just one outcome but several. To link technology and narrative does not yoke two disparate subjects; rather, it recalls an ancient relationship.

Tools are older than written language (perhaps, as the chimpanzee's "fishing stick" suggests, even older than spoken language) and cannot merely be considered passive objects, or "signifieds." Tools are known through the body at least as much as they are understood through the mind. The proper use of kitchen utensils and other tools is handed down primarily through direct observation and imitation of others using them. Technologies are not just objects but also the skills needed to use them. Daily life is saturated with tacit knowledge of tools and machines. Coat hangers, water wheels, and baseball bats are solid and tangible, and we know them through physical experiences of texture, pressure, sight, smell, and sound during use more than through verbal description. The slightly bent form of an American axe handle, when grasped, becomes an extension of the arms. To know such a tool it is not enough merely to look at it: one must sense its balance, swing it, and feel its blade sink into a log. Anyone who has used an axe retains a sense of its heft, the arc of its swing, and its sound. As with a baseball bat or an axe, every tool is known through the body. We develop a feel for it. In contrast, when one is only *looking* at an axe, it becomes a text that can be analyzed and placed in a cultural context. It can be a basis for verifiable statements about its size, shape, and uses, including its incorporation into literature and art. Based on such observations, one can construct a chronology of when it was invented, manufactured, and marketed, and of how people incorporated it into a particular time and place. But "reading" the axe yields a different kind of knowledge than using it.

Telling stories and using tools are hardly identical, but there are similarities. Each involves the organization of sequences, either in words or in mental images. For another investigation it might be crucial to establish whether tools or narratives came first, but for my argument it matters only that they emerged many millennia ago. I do not propose to develop a grand theory of how human consciousness evolved in relation to tools. But the larger temporal framework is a necessary reminder that tools existed long before written texts and that tools have always embodied latent narratives. My definition of technology does not depend on fixing precisely when humans began to use tools, although it is pertinent that they did so thousands of years before anyone developed tools for writing. Cultures always emerge before texts. Long before the advent of writing, every culture had a system of artifacts that evolved together with spoken language. Objects do not define words, or vice-versa; both are needed to construct a cultural world. Only quite late in human development did anyone develop an alphabet, a stylus to mark clay tablets, or a quill adapted for writing on paper. Storytelling was oral for most of human history.

A tool always implies at least one small story. There is a situation; something needs doing. Someone obtains or invents a tool in order to do it—a twisted coat hanger, for example. And afterwards, when the car door is opened, there is a new situation. Admittedly, this is not much of a narrative, taken in the abstract, but to conceive of a tool is to think in time and to imagine change. The existence of a tool also immediately implies that a cultural group has reached a point where it can remember past actions and reproduce them in memory. Tools require the ability to recollect what one has done and to see actions as a sequence in time. To explain what a tool is and how to use it seems to demand narrative. Which came first? This may be a misleading question. It seems more likely that storytelling and toolmaking evolved symbiotically, analogous to the way that oral performances are inseparable from gestures and mimicry.

It is easy to imagine human beings as pre-literate, but it is difficult to imagine them as pre-technological. Most Native American peoples, for example, did not write, but they did develop a wide range of tools, including snowshoes, traps, tents, drums, hatchets, bows, pottery, ovens, bricks, canals, and irrigation systems. All social groups use tools to provide music, shelter, protection, and food, and these devices are inseparable from verbal, visual, and kinetic systems of meaning. Each society both invents tools and selects devices from other cultures to establish its particular technological repertoire of devices.

In Herman Melville's *Moby Dick,* Queequeg, a South Sea harpooner visiting Nantucket, was offered a wheelbarrow to move his belongings from an inn to the dock. But he did not understand how it worked, and so, after putting all his gear into the wheelbarrow he lifted it onto his shoulders. Most travelers have done something that looked equally silly to the natives, for we are all unfamiliar with some local technologies. This is another way of saying that we do not know the many routines and small narratives that underlie everyday life in other societies.

As the evolutionary perspective shows, technology is not something new; it is more ancient than the stone circles at Stonehenge. Great stone blocks, the largest weighing up to 50 tons, rise out of the Salisbury Plain, put precisely into place in roughly 2000 B.C. The stones were not quarried nearby, but transported 20 miles from Marlborough Down. The builders contrived to situate them in a pattern of alignment that still registers the summer solstice and some astronomical events. The builders acquired many technologies before they could construct such a site. Most obviously, they learned to cut, hoist, and transport the stones, which required ropes, levers, rollers, wedges, hammers, and much more. Just as impressive, they observed the heavens, somehow recorded their observations, and designed a monument that embodied their knowledge. They did not leave written records, but Stonehenge stands as an impressive text from their culture, one that we are still learning to read. Transporting and placing the massive stones can only be considered a technological feat. Yet every arrowhead and potshard makes a similar point: that human beings mastered technologies thousands of years ago. Stonehenge suggests the truth of Walter Benjamin's observation that "technology is not the mastery of nature but of the relations between nature and man."[4]

Technologies have been part of human society from as far back as archaeology can take us into the past, but "technology" is not an old word in English. The ancient Greeks had the word "techne," which had to do with skill in the arts. Plato and Plotinus laid out a hierarchy of knowledge that stretched in an ascending scale from the crafts to the sciences, moving from the physical to the intellectual. The technical arts could at best occupy a middle position in this scheme. Aristotle had a "more neutral, simpler and far less value-laden concept of the productive arts."[5] He discussed "techne" in the *Nicomachean Ethics*[6] (book 6, chapters 3 and 4). Using architecture as his example, he defined art as "a rational faculty exercised in making something . . . a productive quality exercised in combination with true reason." "The business of every art," he asserted, "is to bring something into existence." A product of art, in contrast to a product of nature, "has its efficient cause in the maker and not in itself."[7] Such a definition includes such actions as making pottery, building a bridge, and carving a statue. Just as important, Aristotle related the crafts to the sciences, notably through mathematics. In Greek thought as a whole, however, work with the hands was decidedly inferior to philosophical speculation, and "techne" was a more restricted term than the capacious modern term "technology." Perhaps because the term was more focused, classical thinkers realized, Leo Strauss wrote, "that one cannot be distrustful of political or social change without being distrustful of technological change."[8] As

Strauss concluded, they "demanded the strict moral-political supervision of inventions; the good and wise city will determine which inventions are to be made use of and which are to be suppressed."[9]

The Romans valued what we now call technology more highly than the Greeks. In *De Natura Deorum* Cicero praised the human ability to transform the environment and create a "second nature." Other Roman poets praised the construction of roads and the pleasures of a well-built villa. Statius devoted an entire poem to praising technological progress, and Pliny authored prose works with a similar theme.[10] Saint Augustine synthesized Plato and Aristotle with Cicero's appreciation of skilled labor: ". . . there have been discovered and perfected, by the natural genius of man, innumerable arts and skills which minister not only to the necessities of life but also to human enjoyment. And even in those arts where the purposes may seem superfluous, perilous and pernicious, there is exercised an acuteness of intelligence of so high an order that it reveals how richly endowed our human nature is."[11] In contrast, Thomas Aquinas characterized the mechanical arts as merely servile.[12] Some medieval thinkers, notably Albertus Magnus, appreciated iron smelting, the construction of drainage ditches, and the new plowing techniques that minimized erosion. A few drew upon Arabic thought, which presented the crafts as practical science and applied mathematics. Roger Bacon, in his *Communia Mathematica,* imagined flying machines, self-propelled vehicles, submarines, and other conquests of nature. Bacon put so much emphasis on the practical advantages of experiment and construction of useful objects that he "came close to reversing the usual hierarchy of the speculative and useful in medieval thought."[13]

The full expression of a modern attitude toward technology appeared only centuries later, during the Renaissance, notably in

Francis Bacon's *New Atlantis* (1627). Bacon imagined a perfect society whose king was advised by scientists and engineers organized into research groups at an institution called Saloman's House. They could predict the weather, and they had invented refrigeration, submarines, flying machines, loudspeakers, and dazzling medical procedures. Their domination of nature, which had no sinister side effects, satisfied material needs, abolished poverty, and eliminated injustice. This vision helped to inspire others to found the Royal Society.[14] Established in London in 1662, this society institutionalized the belief that science and invention were the engines of progress. The Royal Society proved to be a permanent body, in contrast to earlier, temporary groups that could also be seen as originators of modern research, such as those gathered in Tycho Brahe's astronomical observatory on an island near Copenhagen, or Emperor Rudolf's group of technicians and scientists in Prague.

Today, a large bookstore typically devotes a section to the history of science but scatters books on technological history through many departments, including sociology, cultural studies, women's studies, history, media, anthropology, transportation, and do-it-yourself. The fundamental misconception remains that practical discoveries emerge from pure science and that technology is merely a working out or an application of scientific principles. In fact, for most of human history technology came first; theory came along later and tried to make sense of practical results. A metallurgist at MIT, Cyril Stanley Smith, who helped design the first atomic bombs at Los Alamos, declared: "Technology is more closely related to art than to science—not only materially, because art must somehow involve the selection and manipulation of matter, but conceptually as well, because the technologist, like the artist, must work with unanalyzable complexities."[15] Smith did not mean that these complexities are forever unanalyzable; he meant that at the moment of making something a technologist works within constraints of time, knowledge, funding, and the materials available. It is striking that he advances this argument when discussing the construction of the first atomic bomb, which might seem to be the perfect example of an object whose possibility was deduced from pure science alone. However, Smith is correct to emphasize that the actual design of a bomb required far more than abstract thinking, particularly an ability to work with tools and materials. In fact, one sociologist of science has concluded that, although we cannot turn back the clock and "unlearn" the science that lies behind nuclear weapons, it is conceivable that we will manage to lose or forget the practical skills needed to make them.[16]

As Smith further pointed out, technology's connection to science is generally misunderstood: "Nearly everyone believes, falsely, that technology is applied science. It is becoming so, and rapidly, but through most of history science has arisen from problems posed for intellectual solution by the technician's more intimate experience of the behavior of matter and mechanisms."[17] Often the use of tools and machines has preceded a scientific explanation for how they work or why they fail. Thomas Newcomen, who made the first practical steam engines in Britain, worked as an artist in Aristotle's sense of the term "techne." He conceivably might have heard that a French scientist, Denis Papin, was studying steam and vacuum pumps. However, Newcomen had little formal education and could not have read Papin's account of his experiments, published in Latin (1690) or in French (1695), though he conceivably could have seen a short summary published in English (1697). He never saw Papin's small laboratory apparatus—and even had he seen it, it would not have been a model for his much larger engine. Newcomen's steam engine emerged from the trial and error of practical experiments. Papin's scientific publications were less a basis for inventing a workable steam engine than a theoretical explanation for how a steam engine worked. However, further improvements in the steam engine did call for more scientific knowledge on the part of James Watt and later inventors. Likewise, Thomas Edison built his electrical system without the help of mathematical equations to explain the behavior of electricity. Later, Charles Steinmetz and others developed the theoretical knowledge that was necessary to explain the system mathematically and refine it, but this was after Edison's laboratory group had invented and marketed all the components of the electrical system, including generators, bulbs, sockets, and a wiring system. Science has played a similar role in the refinement of many technologies, including the windmill, the water wheel, the locomotive, the automobile, and the airplane.[18] The Wright Brothers were well-read and gifted bicycle mechanics, and they tested their designs in a wind tunnel of their own invention, but they were not scientists.[19]

If one bears these examples in mind, the emergence of the term "technology" into English from modern Latin in the seventeenth century makes considerable sense. At first, the term was almost exclusively employed to describe a systematic study of one of the arts. A book might be called a "technology" of glassmaking, for example. By the early eighteenth century, a characteristic definition was "a description of the arts, especially the mechanical." The word was seldom used in the United States before 1829, when Jacob Bigelow, a Harvard University professor, published a book titled *Elements of Technology*.[20] As late as the 1840s, almost the only American use of the word was in reference to Bigelow's book.[21] In 1859, the year before he was elected president, Abraham Lincoln gave several versions of a lecture on discoveries and inventions without once using the word.[22] Before 1855, even *Scientific American* scarcely used "technology," which only gradually came into circulation. Instead, people spoke of "the mechanic arts" or the "useful arts" or "invention" or "science" in contexts where they would use "technology" today. A search of prominent American periodicals shows that between 1860 and 1870 "technology" appeared only 149 times, while "invention" occurred 24,957 times. During the nineteenth century the term became embedded in the names of prominent educational institutions such as the Massachusetts Institute of Technology, but it had not yet become common in the discussion of industrialization.[23] "At the time of the Industrial Revolution, and through most of the nineteenth century," Leo Marx writes, "the word *technology* primarily referred to a kind of book; except for a few lexical pioneers, it was not until the turn of [the twentieth] century that sophisticated writers like Thorstein Veblen began to use the word to mean the mechanic arts collectively. But that sense of the word did not gain wide currency until after World War I."[24]

This broader definition owed much to German, which had two terms: "teknologie" and the broader "technik." In the early twentieth century, "technik" was translated into English as "technics."[25] From roughly 1775 until the 1840s, "teknologie" referred to systems of classification for the practical arts, but it was gradually abandoned. During the later nineteenth century, German engineers made "technik" central to their professional self-definition, elaborating a discourse that related the term to philosophy, economics, and high culture. "Technik" meant the totality of tools, machines, systems and processes used in the practical arts and engineering.[26] Both Werner Sombart and Max Weber used the term extensively, influencing Thorstein Veblen and others writing in English. As late as 1934, Lewis Mumford's landmark work *Technics and Civilization* echoed this German usage. However, Mumford also used the term "technology" not in the narrow Germanic sense but in reference to the sum total of systems of machines and techniques that underlie a civilization. In subsequent decades the term "technics" died out in English usage and its capacious meanings were poured into "technology."[27]

Mumford had these larger meanings and the German tradition in mind when he argued that three fundamentally different social and economic systems had succeeded one another in an evolutionary pattern. Each had its own "technological complex." He called these "eotechnic" (before c. 1750), "paleotechnic" (1750–1890), and "neotechnic" (1890 on). Mumford conceived these as overlapping and interpenetrating phases in history, so that their dates were approximate and varied from one nation to another. Each phase relied on a distinctive set of machines, processes, and materials. "Speaking in terms of power and characteristic materials," Mumford wrote, "the eotechnic phase is a water-and-wood complex, the paleotechnic phase is a coal-and-iron complex, and the neotechnic phase is an electricity-and-alloy complex."[28] Although historians no longer use either Mumford's terms or his chronology, the sense that history can be conceived as a sequence of technical systems has become common. Along with this sense of a larger sequence came the realization that machines cannot be understood in isolation. As Mumford put it: "The machine cannot be divorced from its larger social pattern; for it is this pattern that gives it meaning and purpose."[29]

One important part of this pattern that Mumford missed, however, was how thoroughly "technology" was shaped by gender. For example, legal records from the thirteenth and fourteenth centuries show that in rural England women were entirely responsible for producing ale, the most common drink of the peasantry. Men took control of alemaking only when it was commercialized.[30] Similarly, some scholars argue that in the early medieval era European women worked in many trades, but that in early modern times women were gradually displaced by men.[31] Ruth Oldenziel has persuasively extended such arguments into the twentieth century, showing that Western society only relatively recently defined the word "technology" as masculine. Between 1820 and 1910, as the word acquired its present meaning, it acquired male connotations. Before then, "the useful arts" included weaving, potterymaking, sewing, and any other activity that transformed matter for human use. The increasing adoption of the word "technology," therefore, is not simply a measure of the rise of industrialization. It also measures the marginalization of women.[32] In the United States, women were excluded from technical education at the new university-level institutes, such as the Rensselaer Polytechnic Institute (established in 1824) and the Massachusetts Institute of Technology (founded in 1861). Nevertheless, because one could become an engineer on the basis of job experience, there were several thousand female engineers in the United States during the nineteenth century. Likewise, despite many obstacles, there were female inventors. The women's buildings of the great world's fairs in Philadelphia (1876), Chicago (1893), Buffalo (1901), and St. Louis (1904) highlighted women's inventions and their contributions to the useful arts. Furthermore, even though women had been almost entirely excluded from formal engineering education, many worked as technical assistants in laboratories, hospitals, and factories. Engineering was culturally defined as purely masculine, pushing women to the margins or to subordinate positions. Only in recent years have scholars begun to see technology in gendered terms, however, and this realization is not yet widely shared.

Indeed, the meaning of "technology" remained unstable in the second half of the twentieth century, when it evolved into an annoyingly vague abstraction. In a single author's writing, the term could serve as both cause and effect, or as both object and process. The word's meaning was further complicated in the 1990s, when newspapers, stock traders, and bookstores made "technology" a synonym for computers, telephones, and ancillary devices. "Technology" remains an unusually slippery term. It became a part of everyday English little more than 100 years ago. For several hundred years before then, it meant a technical description. Then it gradually became a more abstract term that referred to all the skills, machines, and systems one might study at a technical university. By the middle of the twentieth century, technology had emerged as a comprehensive term for complex systems of machines and techniques.

Indeed, some thinkers began to argue that these systems had a life and a purpose of their own, and no sooner was "technology" in general use than some began to argue for "technological determinism." A single scene in Stanley Kubrick's film *2001* captures the essence of this idea. A primitive ancestor of modern man picks up a bone, uses it as a weapon, then throws it into the air, where it spins, rises, and metamorphoses into a space station. The implications of this scene were obvious: a direct line of inevitable technological development led from the first tools to the conquest of the stars. Should we accept such determinism?

2 | Does Technology Control Us?

Are technologies deterministic?[1] Many people talk as though they are. Students have often told me that the spread of television or the Internet was "inevitable." Likewise, most people find the idea of a modern world without automobiles unimaginable. However, history provides some interesting counterexamples to apparently inevitable technologies. The gun would appear to be the classic case of a weapon that no society could reject once it had been introduced. Yet the Japanese did just that. They adopted guns from Portuguese traders in 1543, learned how to make them, and gradually gave up the bow and the sword. As early as 1575 guns proved decisive in a major battle (Nagoshino), but then the Japanese abandoned them, for what can only be considered cultural reasons. The guns they produced worked well, but they had little symbolic value to warriors, who preferred traditional weapons.[2] The government restricted gun production, but this alone would not be enough to explain Japan's reversion to swords and arrows. Other governments have attempted to restrict gun ownership and use, often with little success. But the Japanese samurai class rejected the new weapon, and the gun disappeared. It re-entered society only after 1853, when Commodore Perry sailed his warships into Japanese waters and forced the country to open itself to the West.

Japan's long, successful rejection of guns is revealing. A society or a group that is able to act without outside interference can abolish a powerful technology. In the United States, the Mennonites and the Amish do not permit any device to be used before they have carefully evaluated its potential impact on the community. For example, they generally resist home telephones and prefer face-to-face communication, although they permit limited use of phones to deal with the outside world. They reject both automobiles and gasoline tractors. Instead, they breed horses and build their own buggies and farm machinery. These choices make the community far more self-sufficient than it would be if each farmer annually spent thousands of dollars on farm machinery, gasoline, and artificial fertilizer, all of which would necessarily come from outside the community. Their leaders decide such matters, rather than leaving each individual to choose in the market. Such practices might seem merely quaint, but they provide a buffer against such things as genetically modified foods or chemical pesticides, and they help to preserve the community. Indeed, the Amish are growing and flourishing. Both the Japanese rejection of the gun and the Amish selective acceptance of modern farming equipment show that communities can make self-conscious technological choices and can resist even very powerful technologies.

Furthermore, these two examples suggest that the belief in determinism paradoxically seems to require a "free market." The belief in technological determinism is widely accepted in individualistic societies that embrace laissez-faire economics. What many people have in mind when they say that television or the Internet was "inevitable" boils down to an assumption that these technologies are so appealing that most consumers, given the chance, will buy them. Historians of technology often reject this view because they are concerned not only with consumers but also with inventors, entrepreneurs, and marketers. They see each new technology not simply as a product to be purchased, but as a part of a larger system. Few historians argue that machines determine history. Instead, they contend that new technologies are shaped by social conditions, prices, traditions, popular attitudes, interest groups, class differences, and government policy.[3]

A surprising number of people, however, including many scholars, speak and write about technologies as though they were deterministic. According to one widely read book, television has "helped change the deferential Negro into the proud Black," has "given women an outside view of their incarceration in the home," and has "weakened visible authorities by destroying the distance and mystery that once enhanced their aura and prestige."[4] These examples suggest that technology has an inexorable logic, that it forces change. But is this the inexorable effect of introducing television into China or the Arab world? In some cases, one might argue, television is strengthening fundamentalism. It simply will not do to assume that the peculiar structure of the American television market is natural. In the United States, television is secular, not religious; private, not public; funded by advertising, not taxation; and a conduit primarily of entertainment, not education. These are cultural choices.

Many have made a similar mistake in writing about the Internet. Nicholas Negroponte declared, in a best-selling book, that "digital technology can be a natural force drawing people into greater world harmony."[5] This is nonsense. No technology is, has been, or will be a "natural force." Nor will any technology by itself break down cultural barriers and bring world peace. Consider the wheel, an invention that most people think of as essential to civilization. Surely the wheel must be an irresistible force, even if the gun and the automobile are not! Much of North Africa, however, let the wheel fall into disuse after the third century A.D., preferring to transport goods by camel. This was a sensible choice. Maintaining roads for wheeled carts and supplying watering sites for horses and oxen was far more expensive, given the terrain and the climate, than opting for the camel, which "can carry more, move faster, and travel further, on less food and water, than an ox," needs "neither roads nor bridges," and is able to "traverse rough ground and ford rivers and streams."[6] In short, societies that have used the wheel may turn away from it. Other civilizations, notably the Mayans and the Aztecs, knew of the wheel but never developed it for practical purposes. They put wheels on toys and ceremonial objects, yet apparently they did not use wheels in construction or transportation. In short, awareness of particular tools or machines does not automatically force a society to adopt them or to keep them.

In *Capitalism and Material Life*, Fernand Braudel rejected technological determinism. Reflecting on how slowly some societies adopt new methods and techniques, he declared: "Technology is only an instrument and man does not always know how to use it."[7] Like Braudel, most specialists in the history of technology do not see new machines as coercive agents dictating social change, and most remain unpersuaded by determinism, though they readily agree that people are often reluctant to give up conveniences. For millennia people lived without electric light or central heating, but during the last 150 years many societies have adopted these technologies and made them part of their building codes. It is now illegal in many places to build or live in a house without indoor plumbing, heating, and electric lighting. In other words, people become enmeshed in a web of technical choices made for them by their ancestors. This is not determinism, though it does suggest why people may come to feel trapped by choices others have made.

Often, adopting a new technology has unintended consequences. Governments build highways to relieve traffic congestion, but better roads may attract more traffic and reduce the use of mass transit as an alternative. Edward Tenner, in his book *Why Things Bite Back,* examines "the revenge of unintended consequences."[8] Among many examples, he notes that computers are expected to improve office efficiency, but in practice people spend enormous amounts of time adjusting to updated software and they suffer eyestrain, back problems, tendonitis, and cumulative trauma disorder.[9] Furthermore, to the extent that computers replace secretaries, white-collar professionals often find themselves doing routine tasks, such as copying and filing documents and stuffing envelopes. Thus, despite many claims made for greater efficiency through computerization, a study by the American Manufacturing Association found that reducing staff raised profits for only 43 percent of the firms that tried it, and 24 percent actually suffered losses, despite the savings on wages. In some cases computerization reduced the time that highly skilled employees had available to perform skilled work. "Their jobs became more diverse in a negative way, including things like printing out letters that their secretaries once did."[10] For some white-collar workers, the computer had the unintended consequence of diminishing their specialization.

In short, rather than assuming that technologies are deterministic, it appears more reasonable to assume that cultural choices shape their uses. While salesmen and promoters like to claim that a new machine is inevitable and urge us to buy it now or risk falling behind competitors, historical experience strongly suggests that the actual usefulness of a new technology is unpredictable.

The idea that mechanical systems are deterministic remains so persistent, however, that a brief review of this tradition is necessary. In the middle of the nineteenth century, most European and American observers saw machines as the motor of change that pushed society toward the future. The phrase "industrial revolution," which gradually came into use after c. 1875, likewise expressed the notion that new technologies were breaking decisively with the past. Early socialists and free-market capitalists agreed on little else, but both saw industrialization as an unfolding of rationality. Even harsh early critics tended to assume that the machine itself was neutral, and focused their attacks on people who misused it. Not until the twentieth century did many argue that technologies might be out of control or inherently dangerous. Technological determinism, which in the nineteenth century often seemed beneficent, appeared more threatening thereafter.

Some Victorians worried that machinery seemed to proliferate more rapidly than the political means to govern it. Without any need of the word "technology," Thomas Carlyle issued a full-scale indictment of industrialization that contained many of the negative meanings that later would be poured into the term. His contemporary Karl Marx saw the mechanization of society as part of an iron law of inevitable historical development.[11] In *The Critique of Political Economy,* Marx argued that "the mode of production of material life determines the general character of the social, political, and spiritual process of life."[12] (Marx did not use the word "technology" in the first edition of *Das Kapital,*[13] though it did appear in later editions. His collaborator, Engels, took up the term "technics" late in life.[14]) Marx argued that industrialization's immediate results were largely negative for the working class. The skilled artisan who once had the satisfaction of making a finished product was subjected to the subdivision of labor. The worker, who once had decided when to work and when to take breaks, lost control of such choices in the new factories. Capital's increasing control of the means of production went along with de-skilling of work and lowering of wages. Industrialization broke the bonds of communities and widened the gaps between social classes. Marx argued that capitalism would collapse not only because it was unjust and immoral, and not only because poverty and inequality would goad the workers to revolt, but also because it would create economic crises of increasing intensity. These crises were not caused by greed or oppression, and they would occur no matter how well meaning capitalists themselves might be. For Marx, the logic of capitalism led to continual investment in better machines and factories, which tied up resources in "fixed capital," leaving less money available for wages ("variable capital"). As investments shifted from labor power to machinery, the amount available for wages and the number of workers employed had to decrease; otherwise the capitalist could not make a profit. This made sense for each individual capitalist, but the overall effect on society when many factories cut total wages and substituted machines for men was a decrease in demand. At the very time when a capitalist had more goods to sell (because he had a new and better production system), fewer people had money to purchase those goods. Thus, Marx argued, efficiency in production flooded the market with goods, but simultaneously the substitution of machines for laborers undermined demand. A crisis was unavoidable. If a capitalist halted production until he had sold off surpluses, he reduced demand still further. If he raised wages to stimulate demand, profits fell. If he sought still greater efficiencies through mergers with rivals, he threw even more workers on the dole, and the imbalance between excessive supply and weak demand became more severe. Marx's analysis posited the inevitable end of capitalism. As greater mechanization produced greater surpluses, it impoverished more workers, causing increasingly severe economic crises because supplies outran demand. Mechanization under capitalism apparently led unavoidably to worker exploitation, social inequality, class warfare, social collapse, and finally revolution.

Marx did not reject technology itself. After the collapse of capitalism, he expected, a succeeding socialist regime would appropriate the means of production and build an egalitarian life of plenty for all. If Marxism made a powerful critique of industrialization that included such concepts as class struggle, worker alienation, de-skilling of artisans, false consciousness, and reification, ultimately it was not hostile to the machine as such. Rather, both Marx and Engels expected that industrialization would provide the basis for a better world. Similarly, Lenin hoped that after the Russian Revolution the technical elite would rationally direct further industrialization and redistribute the wealth it produced. Lenin argued that revolutionary change "should not be confused with the question of the scientifically trained staff of engineers, agronomists and so on." "These gentlemen," he continued, "are working today in obedience to the wishes of the capitalists, and will work even better tomorrow in obedience to the wishes of the armed workers."[15] After the Revolution, the Soviet Union emphasized electrification and mass production. Lenin famously declared that only when the Soviet Union had been completely electrified could it attain full socialism. He vigorously pursued a ten-year plan of building generating plants and incorporated them into a national grid, with the goal of extending electrical service to every home.[16] As this example suggests, Marxists criticized how capitalists used technical systems but not industrialization itself.

The left generally assumed that a society's technologies defined its economic system and social organization. Thus the primitive mill produced feudalism, while the steam engine produced capitalism. They equated mechanization and industrialization with the rational unfolding of history. Evolutionary socialists agreed that technological systems ultimately would become the basis of a utopia, without, however, expecting that violent class conflict and revolution were necessary to attain it. They believed that new technologies would lead to the inevitable decline of capitalism and the emergence of a better economic system. For example, German-born Charles Steinmetz, the leading scientist at General Electric in its first decades, expected socialism to emerge along with a national electrical grid, because it was an inherently interdependent basis for economic reorganization. Electricity could not be stored efficiently and had to be consumed through large distribution systems as soon as it was produced. "The relation between the steam engine as a source of power and the electric motor is thus about the same as the relation between the individualist [capitalist] and the socialist. . . . The one is independent of everything else, is self-contained, the other, the electric motor, is dependent on every other user in the system. . . . The electric power is probably today the most powerful force tending towards co-ordination, that is cooperation [socialism]."[17] Both Marxists and evolutionary socialists embraced not only the machine but also a sense of inevitable historical development based on technological change.

In contrast, Werner Sombart rejected such determinism in *Technik und Kultur,* where he argued that cultures often shaped events more than technologies did. For example, Sombart thought that the failure of cultural and political institutions, and not technological change, accounted for the decline of ancient Rome. Sombart accorded technology an important role in history, particularly in modern times, but he also recognized the importance of culture and institutions. The Chicago School of sociology developed Sombart's ideas in the United States. For example, when William Ogburn wrote about "the influence of invention and discovery," he denied that "mechanical invention is the source of all change" and pointed to "social inventions" such as "the city manager form of government . . . which have had great effects upon social customs. While many social inventions are only remotely connected with mechanical inventions, others appear to be precipitated by" them, such as "the trade union and the tourist camp." Influence could flow in either direction. Social inventions could stimulate technical invention.[18] Ogburn admitted that mechanization had a powerful effect on society, yet he emphasized that "a social change is seldom the result of a single invention." Women's suffrage, for example, was the outcome of a great number of converging forces and influences, including mass production, urbanization, birth control, the adoption of the typewriter, improved education, and the theory of natural rights. Most historical changes were attributable to such a "piling up process." Making the distinction between social invention and technical invention also suggested to Ogburn the notion of a cultural lag. "There is often a delay or lag in the adaptive culture after the material culture has changed, and sometimes these lags are very costly, as was the case with workmen's compensation for industrial accidents."[19] "The more one studies the relationship between mechanical and social inventions," Ogburn concluded, "the more interrelated they seem. Civilization is a complex of interconnections between social institutions and customs on the one hand, and technology and science on the other."[20] Because "the whole interconnected mass is in motion,"[21] it was difficult to establish causation.

The idea that technologies developed more rapidly than society remained attractive to some later theorists. During the 1960s, Marshall McLuhan won a large following as he argued that every major form of communication had reshaped the way people saw their world, causing changes in both public behavior and political institutions. For McLuhan, innovations in communications, notably the printing press, radio, and television, had automatic effects on society. Unlike Ogburn, McLuhan paid little attention to reciprocal effects or social inventions. For McLuhan, not only did the media extend the human sense organs; each new form of a medium disrupted the relationship between the senses. McLuhan argued that the phonetic alphabet intensified the visual function and that literate cultures devalued the other senses—a process that moveable type intensified. Furthermore, McLuhan thought electronic media extended the central nervous system and linked humanity together in a global network. Alvin Toffler reworked such deterministic ideas into *Future Shock,* a best-seller that argued that technological change had accelerated to the point that people scarcely could cope with it. Later, in *The Third Wave,* Toffler argued that a new industrial revolution was being driven by electronics, computers, and the space program.[22] In such studies, the word "impact" suggests that machines inexorably impress change on society.

Although the details of their analyses varied, both McLuhan's arguments and Toffler's were *externalist,* treating new technologies as autonomous forces that compel society to change. The public has an appetite for proclamations that new technologies have beneficent "natural" effects with little government intervention or public planning. Externalist arguments attribute to a technology a dominant place within society, without focusing much on invention or technical details. Externalist studies of "technology transfer" often say little about machines and processes, such as firearms or textile factories, but a great deal about their "impact" on other countries.[23] Externalists usually adopt the point of view of a third-person narrator who stands outside technical processes. They seldom dwell on the (often protracted) difficulties in defining the technological object at the time of its invention and early diffusion. Close analysis—common in the internalist approach to be described in chapter 4—tends to undermine determinism, because it reveals the importance of particular individuals, accidents, chance, and local circumstances.

Determinism is not limited to optimists. Between 1945 and 1970, many of the most pessimistic critics of technology were also determinists. Jacques Ellul paid little attention to the origins of individual inventions, but argued instead that an abstract "Technique" had permeated all aspects of society and had become the new "milieu" that Western societies substituted for Nature. Readers of Ellul's book *The Technological Society*[24] were told that Technique was an autonomous and unrelenting substitution of means for ends. Modern society's vast ensemble of techniques had become self-engendering and had accelerated out of humanity's control: "Technical progress tends to act, not according to an arithmetic, but according to a geometric progression."[25]

Writers on the left found technology equally threatening, and many thought the only possible antidote to be a dramatic shift in consciousness. In *One-Dimensional Man* (1964) and other works, Herbert Marcuse, a Marxist sociologist whose work emerged from the Frankfurt School, attacked the technocratic state in both its capitalist and its socialist formations. He called for "revolutionary consciousness-raising" in preparation for a wholesale rejection of the managed system that everywhere was reducing people to unimaginative cogs in the machine of the state. Marcuse, who became popular with the student movements of the late 1960s, hoped that the "New Left" would spearhead the rejection of the technocratic regime. In *The Making of a Counter Culture* (1969), Theodore Roszak was equally critical but less confrontational, arguing that reform of the technocratic state was impossible. His first chapter, "Technocracy's Children," attacked the mystification of all decision making as it became clothed in the apparently irrefutable statistics and the terminology of technocrats. Western society had become a "technocracy," defined by Roszak as "that society in which those who govern justify themselves by appeal to technical experts who, in turn, justify themselves by appeal to scientific forms of knowledge."[26] Such a technical ideology seemed "ideologically invisible" because its assumed ideals—rationality and efficiency—were accepted without discussion both in the communist East and the capitalist West. To resist technocracy, a de-technologized consciousness was needed, which Roszak sought through a combination of Zen Buddhism, post-Freudian psychology, and the construction of alternative grassroots institutions, such as those in the emerging hippie movement.[27]

As student radicalism faded during the 1970s, social revolution seemed less probable than technological domination, notably as analyzed in the work of Michel Foucault. He treated technology as the material expression of an overarching discourse that structured individual consciousness and shaped institutions, notably hospitals, asylums, and prisons.[28] In contrast to Marx, Foucault's theory did not conceive of an economic or a technical "base" that drove changes in the social "superstructure." Rather, Foucault saw history as the exfoliation of patterns of ideas and structures ("epistemes"), which were expressed in art, in architecture, in classification systems, in social relations, and in all other aspects of the cultural discourse at a given historical moment. The epistemes did not evolve from one discursive system to the next but rather were separated by ruptures, or breaks in continuity. When a new discourse emerged, it did not build upon previous systems. Rather, as a sympathetic critic summarized, "a new knowledge begins, it is unrelated to previous knowledge."[29] Foucault conceived history as a series of internally coherent epistemological systems, each built upon different premises. The individual author, inventor, or citizen was not the master of his or her fate but rather was penetrated and defined by discourses. Each was caught within, scarcely aware of, and ultimately articulated by structures of knowledge and power that were deployed and naturalized throughout society. In the modern episteme, Foucault was concerned with how power became anonymous and embedded in bureaucracies, making hierarchical surveillance a social norm. His determinism was far more comprehensive than that of most previous thinkers.

Foucault, and later the postmodernist Francois Lyotard, authored academic best-sellers of the 1970s and the 1980s, but their grand deterministic theories found little favor among historians of technology, whose research showed considerable evidence of human agency in the creation, dissemination, and use of new technologies. Leo Marx declared that postmodern theorists in effect ratify "the idea of the domination of life by large technological systems" and promote a "shrunken sense of human agency."[30] The most sweeping rejection of technological determinism came from Marx's student Langdon Winner in *Autonomous Technology*, a book Winner said he had written in a spirit of "epistemological Luddism."[31] In dismantling deterministic ideologies, Winner made it easier to think of technologies as socially shaped, or constructed. Winner also emphasized Karl Marx's more flexible views of technology in his earlier works. In

The German Ideology (1846), Winner comments, "human beings do not stand at the mercy of a great deterministic punch press that cranks out precisely tailored persons at a certain rate during a given historical period. Instead, the situation Marx describes is one in which individuals are actively involved in the daily creation and recreation, production and reproduction of the world in which they live."[32] While Marx's labor theory of value might seem to suggest rigid determinism, Winner argues that his work as a whole does not support such a view.

Technological determinism lacks a coherent philosophical tradition, although it remains popular. A variety of thinkers on both the right and the left have put forward theories of technological determinism, but the majority of historians of technology have not found them useful. As the following two chapters will show, deterministic conceptions of technology seem misguided when one looks closely at the invention, the development, and the marketing of individual devices.

3 Is Technology Predictable?

If technologies are not deterministic, then neither their emergence nor their social effects should be predictable. To consider this proposition in detail, we can divide technological prognostication into three parts: prediction, forecasting, and projection. We predict the unknown, forecast possibilities, and project probabilities. These three terms correspond to the division common in business studies of innovation, between what James Utterbeck terms "*invention* (ideas or concepts for new products and processes), *innovation* (reduction of an idea to the first use or sale) and *diffusion* of technologies (their widespread use in the market)."[1]

Prediction concerns inventions that are fundamentally new devices. This is a more restrictive definition than the US Patent Office's sense of "invention," for that also includes "innovation," treated here as a separate category. What is the distinction? The incandescent electric light was an invention; new kinds of filaments were innovations. The telephone was an invention, but the successive improvements in its operation were innovations. Inventions are fundamental breakthroughs, and there have been relatively few. In communications, they would include the telephone, the electric light, radio, television, the mainframe computer, the personal computer, and the Internet. While prediction

Table 3.1

Form of prognostication	Persons typically involved	Their focus	Their time frame
Prediction	Inventors, utopian writers	Breakthrough inventions	Long term
Forecasting	Engineers, entrepreneurs	Innovations	Less than 10 years
Projection	Designers, marketers	New models	Less than 3 years

concerns such inventions, forecasting concerns innovations, which are far more numerous. Innovations are improvements and accessories to systems that emerged from inventions. The third term, "projection," which I will discuss only briefly, concerns the future sales, profits, market share, and so forth of new models of established technologies.

Prediction, forecasting, and projection typically involve different professionals working within different time frames. (See table 3.1.) These distinctions are not merely a matter of semantic convenience. If one looks at the time frames involved, prediction deals with the long term or even indefinite periods, whereas forecasting focuses on immediate choices about getting a new device perfected and into production. Those making projections must work within the shortest time frame, because they deal with new (often annual) models of devices that compete in the market.

Who is centrally involved in prognostication depends on which category one is dealing with. Inventors, futurologists, and some academics predict or debunk the possibility of fundamental breakthroughs. Once a workable device exists, however, venture capitalists, engineers, and consultants busy themselves with forecasting its possibilities. If a device is widely accepted, designers and marketers take a central role in projecting and extrapolating what new styles and models consumers will buy. In view of the differences in actors and in time frames, there are considerable differences in the aesthetics of invention, innovation, and product development, emphasizing, respectively, technical elegance, functionalism, and beauty.[2]

On television one mostly hears forecasting and projection, not prediction. For example, in 1998 a "technology guru" on the Cable News Network announced that voice recognition would be the "next big thing" in computers because keyboards could then be done away with, and small computers capable of responding to verbal commands would be embedded in useful objects everywhere.[3] Machine speech recognition was already used by telephone companies by that time; its possible extension and development to replace computer keyboards was forecast. Eight years later, voice recognition seems to have spread more slowly than that "guru" expected.

All technological predictions and forecasts are in essence little narratives about the future. They are not full-scale narratives of utopia, but they are usually presented as stories about a better world to come. The most successful present an innovation as not just desirable but inevitable. Public-relations people are well aware that such stories can become self-fulfilling when investors and consumers believe them. As the consultant and critic John Perry Barlow once put it, "the best way to invent the future is to predict it—if you can get enough people to believe your prediction, that is."[4]

Selling stories of the wonders to come has been popular at least since the Chicago World's Columbian Exposition of 1893,[5] and they have become the stock in trade of investment newsletters, some technical magazines, and certain educational television programs. To put this another way, inventors and corporate research departments create not only products but also compelling narratives about how these new devices will fit into everyday life. They need to do this to get venture capital, and companies need to market such scenarios to get a return on investment.

Yet accurate prediction is difficult, even for experts. George Wise, a historian who worked for years at the General Electric research labs in Schenectady, wrote his doctoral thesis on how well scientists, inventors, and sociologists predicted the future between 1890 and 1940. Examining 1,500 published predictions, he found that only one-third proved correct, while one-third were wrong and another one-third were still unproved. They used many methods, including intuition, analogy, extrapolation, studying leading indicators, and deduction, but all were of roughly equal accuracy.[6] The technical experts, he found, performed only slightly better than others. In short, technological predictions, whoever made them and whatever method was employed, proved no more accurate than flipping a coin.

If prediction has proved extremely difficult, what about forecasting? That ought to be easier, because it deals with already invented technologies and builds on existing trends. Anyone interested in computers has heard of Moore's Law, formulated in 1965, which predicted, quite accurately, that computer memory would double roughly every 18 to 24 months.[7] (Note, however, that this may have been a self-fulfilling prophecy, because it established a benchmark for development in the computer industry.) Yet for every such success there are famous failures of forecasting. No demographer saw the United States' post-World War II baby boom coming. American birth rates had fallen steadily for more than 100 years, and demographers were surprised when the decline did not continue. In the 1960s a great many sociologists projected that automation would reduce the average American's work week to less than 25 hours by the century's end. Instead, the average American today is working more hours than in 1968.[8] Paul Ehrlich, in *The Population Bomb,* predicted in the early 1970s that it was already too late to save India from starvation.[9] He did not foresee the tremendous increases in agricultural productivity. Social trends are difficult to anticipate. General forecasting is risky, failure common.

Technological forecasting is no easier. In 1900, few investors forecast that the new automobiles would replace trolley cars.[10] Trolley service had grown tremendously in the previous decade, and it was expanding into long-distance competition with the railroad. The automobile was still a rich man's toy, and no one anticipated the emergence or the tremendous productivity of the automotive assembly line. In the 1930s, when only one in a hundred people had actually been up in an airplane, a majority of Americans mistakenly expected that soon every family would have one.[11] In 1954, Chairman Lewis Strauss of the US Atomic Energy Commission told the National Association of Science Writers that their children would enjoy "electrical energy too cheap to meter."[12] IBM, thinking that mainframes would always be the core of the computer business, waited seven years before competing directly with Digital Computer's minicomputers.[13] Later, Apple mistakenly thought there was a market for its Newton, an early personal digital assistant that had good handwriting recognition but proved too large and too expensive for most consumers. The experts at Microsoft did not foresee the sudden emergence of the World Wide Web, and were slow to compete with Netscape when it appeared. These were all failures of forecasting.

Projection might be expected to work reasonably well when the economy is stable. The total demand for most items will be stable, and extrapolations based on growth rates *may* prove accurate. But a stable market is full of competing products, and full of expanding and contracting firms. In the 1950s, Ford thought there was a market for the Edsel. Furthermore, business conditions are seldom stable for long. In the 1960s, American utility companies expected growth in the consumption of electricity to double every ten years, as it had done for decades. The utility companies did not foresee the energy crises of the 1970s, which would trigger a move toward conservation.[14] The energy crisis likewise caught American automakers unprepared; they had projected continued demand for large cars, and they had few small, energy-efficient vehicles for sale.

As these examples suggest, any trend that seems obvious, and any pattern that seems persistent, may be destabilized by changes in the economy, changes in technology, or some combination of social and technical factors. As the mathematician John Paulos put it, "futurists such as John Naisbitt and Alvin Toffler attempt to 'add up' the causes and effects of countless local stories in order to identify and project trends." But "interactions among the various trends are commonly ignored, and unexpected developments, by definition, are not taken into account. As with weather forecasters, the farther ahead they predict, the less perspicacious they become."[15]

It is not just futurists who stumble. Fundamental innovations almost always seem to come from outside the established market leaders, who suffer from "path dependency." Established firms are usually too committed to a particular conception of what their product is. This commitment is embedded in its manufacturing process and endemic in the thinking of its managers. When a major innovation appears, a leading firm understands the technology, but remains committed to its product and its production system. The case of IBM and the personal computer is a good example. At first IBM did not take the threat seriously enough, and competitors had the market for personal computers to themselves for at least four years before IBM entered the field. IBM then was clever enough to license others to manufacture its system, making it the standard, but it had to share the market with many other firms. In 2005, after 25 years, it withdrew from the market.

In most cases, when an innovation such as the personal computer appears, established industries redouble their commitment to the traditional product that has made them the market leader. They make incremental improvements in manufacturing, and yet they lose market share to the invader. This occurs even in fast-changing electronic industries, where innovations come so frequently that there is little time for routines and habits to blind participants to the advantages of the next change. Utterback cites a comprehensive study of the manufacturers that supply semiconductor firms with photolithographic alignment machines. During the invention and development of five distinct generations of such machines, in no case did the market leader at one stage retain its top position at the next.[16] A production system seems to gain such a powerful hold inside a firm that it seldom can move swiftly enough to adopt innovations.[17]

Another reason that forecasts and predictions are so hard to make is that consumers, not scientists, often discover what is "the next big thing." Most new technologies are market-driven. Viagra was not developed as a sexual stimulant, but the college students who served as guinea pigs discovered what consumers would like about it. This general point can be put negatively: Just because something is technologically feasible, don't expect the public to rush out and buy it. Consumers must want the product. There were many mistaken investments in machines that worked but which the public didn't want. The classic case may be AT&T's Picture Phone.[18] It was technologically feasible, and it was promoted at the New York World's Fair of 1964. But aiming immediately at the mass market, rather than starting more slowly with a niche market, proved a miscalculation. Few bought it, partly because they resisted its high price but also because they feared a visual invasion of their privacy and because they did not understand its potential as a data-display terminal. Though some apparently reasonable technologies fail to sell, people may nonetheless flock to "unreasonable" devices, such as Japanese electronic pets.

Histories of new machines tend to focus on the process of invention and to suggest that the market is driven by research and development. This is usually not so, even in the case of inventions that in retrospect clearly were fundamental to contemporary society: the telegraph, the telephone, the phonograph, the personal computer. When such things first appear, creating demand is more difficult than creating supply. At first, Samuel Morse had trouble convincing anyone to invest in his telegraph. He spent five years "lecturing, lobbying, and negotiating" before he convinced the US Congress to pay for the construction of the first substantial telegraph line, which ran from Washington to Baltimore. Even after it was operating, he had difficulty finding customers interested in using it.[19] Likewise, Alexander Graham Bell could not find an investor to buy his patent on the telephone, and so he reluctantly decided to market it himself.[20] Thomas Edison found few commercial applications for his phonograph, despite the sensational publicity surrounding its discovery.[21] He and his assistants had the following commercial ideas a month after the phonograph was first shown to the world: to make a speaking doll and other toys, to manufacture speaking "clocks . . . to call the hour etc., for advertisements, for calling out directions automatically, delivering lectures, explaining the way," and, almost as an afterthought at the end of the list, "as a musical instrument."[22] In the mid 1970s, a prototype personal computer, when first shown to a group of MIT professors, seemed rather uninteresting to them.[23] They could think of few uses for it, and they suggested that perhaps it would be most useful to shut-ins.

In short, the telegraph, the telephone, the phonograph, and the personal computer, surely four of the most important inventions in the history of communications, were initially understood as curiosities.[24] Their commercial value was not immediately clear. It took both investors and the public time to discover what they could use them for. Eventually large corporations would manufacture each of these inventions, and each became the basis for an international form of communication. As people became familiar with these four technologies, they built them into daily life. Barlow argues that the public's slow response time is generational: ". . . it takes about thirty years for anything really new to arise from an invention, because that's how long it takes for enough of the old and wary to die."[25]

People need time to understand fundamental inventions, which is why they spread slowly; in contrast, innovations are easier to understand and proliferate rapidly. The few fundamental inventions become the bases for entirely new systems, but most innovations plug into an existing system. Once the electrical grid, the telephone network, or the World Wide Web had been built, new application technologies or innovations proliferated. For example, as the electrical grid spread across the United States, small manufacturers rushed in with a stunning array of new products—electrified cigar lighters, model trains, Christmas tree lights, musical toilet-paper dispensers, and shaving cream warmers, as well as toasters, irons, refrigerators, and washing machines. As electric devices proliferated, the large manufacturers Westinghouse and General Electric, like the computer hardware makers of today, soon found it impossible to compete in every area. Once several million PCs and Macs were in place, programmers created the software equivalent of the earlier appliances, with thousands of programs to compose music, calculate income tax, make architectural drawings, encrypt messages, write novels, and so on. Ordinary consumers played a leading role by encouraging such innovations. They drove the rapid growth in sales of scanners, color printers, high-speed modems, external cartridge drives, and software that sends and receives snapshots and short videos.[26]

Selling the basic hardware for a communication system often ceases to be as profitable as selling software and services.[27] People now spend far more money on things that use electricity than on the electricity itself, and this disproportion has been increasing since the 1920s.[28] Something similar happened with the telephone. AT&T began with an absolute monopoly and expanded slowly during the period when no one could compete. During the 1890s, however, AT&T's patent protection ran out, competitors appeared, the market doubled and redoubled in size, and the cost of telephone calls began to drop.[29] The intensity of telephone use and the number of applications was still increasing 100 years later. Where once the telephone bill reflected a simple transaction between a customer and the phone company, now the technology of the telephone is the basis for a wide range of commercial relations that includes toll-free calls to businesses, e-mail, faxes, and SMS messages. Telephones enable people not only to speak to one another, but also to send photographs, texts, news, and videos. As with the electrical system, the telephone provided the infrastructure, or even the main platform, for many unanticipated businesses. The recent proliferation of communication technologies interweaves and connects the electrical grid, the telephone, the television, the personal computer, and the Internet. The synergy of this mix of networked systems makes possible a particularly rich period of innovation. Many possibilities are latent or only partially developed, and that puts a premium on forecasting for the near future.

In this dynamic market, the best design does not always win. Even if someone can accurately foresee the coming of a new technology or an innovation, no one can be certain what design will prove most popular. Perhaps failure was obvious for the air-conditioned bed, the illuminated lawn sprinkler, and the electrically sterilized toilet seat, all marketed in the 1930s,[30] but it was by no means obvious that Sony's Betamax, the technically better machine, would lose out to VHS in the home video market. Marketing, not technological excellence, proved crucial. Sony decided not to share its system with others and expected to reap all the rewards. Its rival, JVC, allied itself with other manufacturers and licensed them to co-produce its VHS system. Consumers decided that more films were likely to be available in the VHS format because a consortium of companies stood behind it, and Betamax gradually lost out.[31] Perhaps the most familiar recent example of a superior machine capturing only a small part of the market is that of Apple's Macintosh computers. Here again a decision to "go it alone" appears to have been a decisive mistake.[32] A somewhat different example is the case of FM radio, which is better for short-distance transmission than AM. It languished virtually unused for a generation because RCA discouraged its use while promoting its already well-established AM network.[33]

Consider an example from the electrical industry: district heating vs. individual home heating. A hundred years ago, most power stations were near town centers, and they routinely marketed excess steam for the heating of apartment blocks, office buildings, and department stores. Since then district heating has failed to capture much of the American market,[34] although in Scandinavia district heating is popular because it saves energy, lowers pollution levels, and reduces the cost of home heating. District heating was also widespread and apparently worked well in the former Soviet Union, but the plants are now often shut down for "cleaning," especially during the summer, leaving apartments without hot water. American social values emphasize individualized technologies. Every house has its own heating system, even though this is a wasteful and inefficient choice. If the market to some extent shapes technologies, the market in turn is inflected by cultural values.

Even if one can predict which new technologies are possible and forecast which designs will thrive in the market, people may fail to foresee how they will be used. Edison invented the phonograph, but he thought his invention primarily would aid businessmen, who could use it to dictate letters, and he did not focus on music and entertainment even as late as 1890.[35] As a result, competitors grabbed a considerable share of the market, and their system of a flat record on a turntable won out over his turning cylinders. Another example: Between 1900 and 1920 the new technology of radio was perceived by government and industry as an improved telegraph that needed no wires. They expected it to be used for point-to-point communications. When radio stations emerged after World War I as consumer-driven phenomena, the electrical corporations were caught off guard, but they quickly moved into the new market.[36] The public used both the phonograph and the radio less for work than for fun.[37] Likewise, many children use personal computers less to write papers and pursue education than to play computer games and visit strange websites. These activities may or may not be educational; my point is that they were unanticipated and consumer driven.

Another example of unanticipated use is the higher-than-expected consumption of electricity by refrigerators, which so puzzled a California utility company that it hired anthropologists to find out what was going on.[38] They discovered that families used the refrigerator for much more than food storage. It was also a place to hide money in fake cabbages, to protect photographic film, to give nylon stockings longer life, to allow pet snakes to hibernate, and to preserve drugs. At times people opened the refrigerator and gazed in without clear intentions, mentally foraging, trying to decide if they were hungry, often removing nothing before they closed the door again. The anthropologists concluded that the refrigerator, and by extension any tool, "enters into the determination of its own utilities, suggesting new ideas for its own definition . . . and . . . threatens to take on altogether new identities. . . ." The Internet offers a final, stunning example of this principle. Only military planners and scientists initially used this communication system. They developed a decentralized design so that messages could not easily be knocked out by power failures, downed computers, or a war. But this same feature made it difficult to monitor and control the Internet. The developers did not imagine such things as Amazon.com, pornography on the net, downloading digitized music to a personal computer, or most of the other things people today use the Internet for. In short, when we review the history of the phonograph, the radio, the refrigerator, and the Internet, technologies conceived for one clearly defined use have acquired other, unexpected uses over time. Engineers and designers tend to think new devices will serve a narrow range of functions, while the public has a wide range of intentions and desires and usually brings far more imagination to new technologies than those who first market them.

Furthermore, a technology's symbolic meanings may determine its uses. Too often we think of technologies in purely functional terms. However, even so prosaic a device as the electric light bulb had powerful symbolic meanings and associations at its inception. Edison's practical incandescent light of 1879 was preceded by many forms of "impractical" electric lighting in theaters, where it was used for dramatic effects. A generation before Edison's light bulb even began to reach most homes (after 1910), it was appropriated by the wealthy for conspicuous consumption, used to illuminate public monuments and skyscrapers, and put into electrical signs. As a result, by 1903 American cities were far more brightly lighted than their European counterparts: Chicago, New York, and Boston had three to five times as many electric lights per inhabitant as Paris, London, or Berlin.[39] Intensive electric lighting of American downtowns far exceeded the requirements of safety. The Great White Way and its huge signs had become a national landmark by 1910, and postcards and photographs of illuminated city skylines became common across the United States. In New York, during World War I when wartime energy saving darkened Times Square, the citizens complained that the city seemed "unnatural." People demanded that the giant advertising signs be turned on again, and they soon were, with new slogans selling war bonds.[40]

This intensive use of lighting in the United States was in no sense a necessity, and the European preference for less electric advertising was not temporary or the expression of a "cultural lag." Many European communities still resist electric signs and spectacular advertising displays. At the 1994 Winter Olympics in Norway, the city council of Lillehammer refused Coca-Cola and other sponsors the right to erect illuminated signs. On the city's streets only wooden and metal signs were permitted. No neon or transparent plastic was allowed. Levels and methods of lighting vary from culture to culture, and what is considered normal or necessary in the United States may seem to be a violation of tradition elsewhere.[41]

The preceding survey shows that, far from being deterministic, technologies are unpredictable. A fundamentally new invention often has no immediate impact; people need time to find out how they want to use it. Indeed, the best technologies at times fail to win acceptance. Furthermore, the meanings and uses people give to technologies are often unexpected and non-utilitarian. Economics does not always explain what is selected, how it is used, or what it means. From their inception, technologies have symbolic meanings and non-utilitarian attractions.

A technology is not merely a system of machines with certain functions; rather, it is an expression of a social world. Electricity, the telephone, radio, television, the computer, and the Internet are not implacable forces moving through history, but social processes that vary from one time period to another and from one culture to another. These technologies were not "things" that came from outside society and had an "impact"; rather, each was an internal development shaped by its social context. No technology exists in isolation. Each is an open-ended set of problems and possibilities. Each technology is an extension of human lives: someone makes it, someone owns it, some oppose it, many use it, and all interpret it. Because of the multiplicity of actors, the meanings of technology are diverse. This insight is useful for considering how historians understand technology (chapter 4) and for looking into the relationship between technology and cultural diversity (chapter 5).

The previous two chapters suggest that one must reject technological determinism and admit that the invention and the diffusion of technologies are not predictable. What is the alternative?
Historians in the field give roughly equal weight to technical,
social, economic, and political factors. Their case studies suggest
that artifacts emerge as the expressions of social forces, personal
needs, technical limits, markets, and political considerations.
They often find that both the meanings and the design of an artifact are flexible, varying from one culture to another, and from
one time period to another. Indeed, Henry Petroski, one of the
most widely read experts on design, argues that there is no such
thing as perfect form: "Designing anything, from a fence to a factory, involves satisfying constraints, making choices, containing
costs, and accepting compromises."[1] Technologies are social constructions. Historians of technology have also generally agreed
that after initial invention comes an equally important stage of
"development." Indeed, since the late 1960s the study of development has been at the center of much work in the field, for example
in studies of Nicholas Otto's internal-combustion engine or the
development of the diesel engine.[2] Nathan Rosenberg, a leading
economic historian, emphasizes that for every new product or

production technique "there is a long adjustment process during which the invention is improved, bugs ironed out, the technique modified to suit the specific needs of users, and the 'tooling up' and numerous adaptations made so that the new product (process) can not only be produced but can be produced at low cost." Indeed, during this "shakedown period" of early production some feasible inventions are abandoned as unprofitable.[3] As the study of development has increased, the heroic "lone inventor" has largely disappeared from scholarship. Anthony F. C. Wallace, a senior historian in the field, declared: "We shall view technology as a social product and shall not be over much interested in the priority claims of individual inventors, for the actual course of work that leads to the conception and use of new technology *always* involves a group that has worked for a considerable period of time on the basic idea before success is achieved."[4]

Variation in design continues during early stages of development, until one design meets with wide approval. Once a particular design is widely accepted, however, variation in form gives way to innovation in production. Take the bicycle as an example. The earliest bicycles (high-wheelers) were handmade and cost an ordinary worker a year's wages. Only the well-to-do could afford them. Most riders were young and male. The danger of toppling from a high-wheeler gave bicycling a macho aura. For more than a generation, low-wheel bikes were for women, clergymen, and old people. Tremendous experimentation took place as inventors changed the size of the wheels, made three-wheelers, moved the larger wheel from the front to the back, and tried out various materials, including wood and steel frames. They developed different drive and braking systems, tried various shapes and positions for the handlebars, and created accessories such as lights and panniers. Dunlop developed air-filled rubber tires that together

with a padded seat reduced discomfort from bumps and vibra-
tions. At first, professional racers derided these "balloon" tires as
being for sissies. However, bicycle design reached closure in the
1890s, when low-wheel models with front and back wheels of
equal size and fitted with Dunlop tires proved to be the fastest on
the racetrack.

Once the low-wheel "safety bicycle" had become accepted as
the standard, manufacturing changed. The leading producers of
high-wheelers, such as the Pope Company, had prided themselves
on durable construction by skilled artisans, who adjusted the
wheel size of each cycle to match the length of a customer' legs.
In the 1880s such bicycles cost $300 or more, well beyond the
reach of the average consumer. In the 1890s, however, mass producers such as Schwinn made bicycles with stamped and welded
frames. They were of lesser quality, but priced as low as $50. By
1910 a used bicycle in working order could be had for $15, and
ownership had spread to all segments of society. The bicycle had
ceased to be a toy for the wealthy and had become a common form
of transportation and recreation for millions. The military had
adapted it to troop transport, delivery services had thousands of
bicycle messenger boys, and bicycle racing had emerged as a
professional sport.

The social significance and use of the bicycle was not technologically determined. For example, from the beginning some
women adopted the bicycle, and during the era of the high-
wheeler some joined the popular bicycle clubs. In 1888, eighteen
women were members of a Philadelphia club, and one of them
won the "Captain's Cup," awarded annually to the member who
covered the most miles (in this case, 3,304¼). However, women on
wheels met opposition. Some physicians declared that the bicycle
promoted immodesty in women and harmed their reproductive

organs. Moralists thought women on bicycles were indecent because they wore shorter skirts to ride them, and worried that women would find straddling the seat sexually stimulating. The bicycle craze helped kill the bustle and the corset and encouraged "common-sense dressing." Many in the women's suffrage movement adopted bicycles. In 1896, Susan B. Anthony declared: "Bicycling . . . has done more to emancipate women than anything else in the world. I stand and rejoice every time I see a woman ride by on a wheel. It gives women a feeling of freedom and self-reliance."[5] The women's movement embraced the bicycle, and its democratization became part of their drive for social equality.

Outside the United States, the bicycle persisted much longer as an important form of transportation. In the Netherlands and in Denmark, bicycles were more common than automobiles were until the early 1960s. In those countries, major roads have special lanes and special traffic signals for bicyclists, and government programs encourage citizens to use bicycles instead of automobiles. But most Western societies have chosen the automobile as the primary mode of transportation instead, and even in Denmark and the Netherlands bicyclists are not as numerous as they were a generation ago. Despite the bicycle's head start on the automobile, in most societies only the automobile seemed to achieve what Thomas Hughes calls "technological momentum."

Hughes argues that technical systems are not infinitely malleable. If technologies such as the bicycle or the automobile are not independent forces shaping history, they can still exercise a "soft determinism" once they are in place. "Technological momentum" is a particularly useful concept for understanding large-scale systems, such as the electric grid, the railway, or the automobile. In *Networks of Power,* Hughes examines five stages of system development for the electrical grid, and these stages can apply to other inventions as well. In the case of electrification, the sequence began in the 1870s with invention and early development in a few locations (1875–1882). That was followed by technology transfer to other regions (1882–1890). With successful transfer came growth (1890–) and the development of subsidiary infrastructures of production, education, and consumption, leading to technological momentum (after c. 1900) as electricity became a standard source of light, heat, and power. In the mature stage (after. c. 1910), the problems faced by management required financiers and consulting engineers.[6]

"Technological momentum" is not inherent in any technological system when first deployed. It arises as a consequence of early development and successful entrepreneurship, and it emerges at the culmination of a period of growth. The bicycle had such momentum in Denmark and the Netherlands from 1920 until the 1960s, with the result that a system of paved trails and cycling lanes were embedded in the infrastructure before the automobile achieved momentum. In the United States, the automobile became the center of a socio-technical system more quickly and achieved momentum a generation earlier. Only some systems achieve "technological momentum," which Hughes has also applied to analysis of nitrogen fixation systems and atomic energy.[7] The concept seems particularly useful for understanding large systems. These have some flexibility when being defined in their initial phases. But as technical specifications are established and widely adopted, and as a system comes to employ a bureaucracy and thousands of workers, it becomes less responsive to outside pressures. Hughes provided an example in *American Genesis:* ". . . the inertia of the system producing explosives for nuclear weapons arises from the involvement of numerous military, industrial, university, and other organizations, as well as from the commitment of hundreds of thousands of persons whose skills and employment are dependent on the system."[8] Similarly, at the end of the nineteenth century, once the width of railway tracks had been made uniform and several thousand miles were laid out, once bridges and grade crossings were designed with rail cars of certain dimensions in mind, it was expensive and impractical to reconfigure a railway system.

Hughes makes clear when discussing "inertia" that the concept is not only technical but also cultural and institutional. A society may choose to adopt either direct current or alternating current, or to use 110 volts, or 220 volts, or some other voltage, but a generation after these choices have been made it is costly and difficult to undo such a decision. Hundreds of appliance makers, thousands of electricians, and millions of homeowners have made a financial commitment to these technical standards. Furthermore, people become accustomed to particular standards and soon begin to regard them as natural. Once built, an electrical grid is "less shaped by and more the shaper of its environment."[9] This may sound deterministic, but it is not entirely so, for people decided to build the grid and selected its specifications and components. To later generations, however, such technical systems seem to be deterministic.[10]

The US electrical system achieved technological momentum around 1900. By that time, it was "reinforced with a cultural context, and interacting in a systematic way with the elements of that context," and "like high momentum matter [it] tended in time to resist changes in the direction of its development."[11] From 1900 on, growth was relentless and not easily deflected by contingencies. The electrical system was far more than machines themselves; it included utility companies, research laboratories,

regulatory agencies, and educational institutions, constituting what Hughes calls a "sociotechnical system." It had high momentum, force, and direction because of its "institutionally structured nature, heavy capital investments, supportive legislation, and the commitment of know-how and experience."[12] Similarly, the automobile achieved technological momentum not as an isolated machine, but as part of a system that included road building, driver education programs, gas stations, repair shops, manufacturers of spare parts, and new forms of land use that spread out the population into suburbs that, practically speaking, were accessible only to cars and trucks.

The concept of technological momentum provides a way to understand how large systems exercise a "soft determinism" once they are in place. Once a society chooses the automobile (rather than the bicycle supplemented by mass transit) as its preferred system of urban transportation, it is difficult to undo such a decision. The technological momentum of a system is not simply a matter of expense, although the cost of building highways, bicycle lanes, or railroad tracks is important. Ultimately, the momentum of a society's transport system is embodied in the different kinds of cities and suburbs fostered by each form of transportation. Relying on bicycles and streetcars has kept Amsterdam densely populated, which in turn means that relatively few kilometers of streetcar line can efficiently serve the population. If the Dutch were to decide to rely more on the automobile, they would have to rip apart a tightly woven urban fabric of row houses, canals, and small businesses. In contrast, cities such as Houston, Phoenix, and Los Angeles sprawl over larger areas, with more than half the land area devoted to roads, parking lots, garages, gas stations, and other spaces for automobiles. Such a commitment to the automobile has resulted in massive infrastructure investments that make it impractically expensive to shift to mass transit, not least because the houses are so far apart. In the United States the automobile now is "less shaped by and more the shaper of its environment."[13] Hughes's idea of "technological momentum" is far less deterministic than externalist theories, and provides a useful way to think about how large socio-technical systems operate in society.

Most historians of technology are either contextualists or internalists.[14] These are not so much opposed schools of thought as different emphases. Internalists reconstruct the history of machines and processes focusing on the role of the inventor, laboratory practices, and the state of scientific knowledge at a particular time. They chart the sequence that leads from one physical object to the next. The internalist approach has some affinities with art history,[15] but it grew out of the history of science. A five-volume *History of Technology* published in the 1950s detailed the histories of industrial chemicals, textile machinery, steelmaking, electric lighting and generating systems, and so forth.[16] Internalists establish a bedrock of facts about individual inventors, their competition, their technical difficulties, and their solutions to particular problems.

An internalist may be a feminist working on Madame Curie or a railroad historian interested in how different kinds of boxcars developed. Tracy Kidder's best-selling book *The Soul of a New Machine* is an example.[17] Kidder spent months observing a team that was inventing a new computer, charting the work process, the pressures from management for rapid results, the continual advances in electronics that made it hard to know when to freeze the design of the machine, and the step-by-step developments that led to a final product. The book ends shortly after the company presents the new computer to the public at a press conference. It does not examine the consumer's response to the computer or tell the reader much about its sales success. The internalist writes from the point of view of an insider who looks over an inventor's shoulder. Such studies, whether of the light bulb, the computer, or the atom bomb, culminate at the moment when the new device is ready for use.

If many non-specialists believe that necessity is the mother of invention, internalists usually find that creativity is by no means assured or automatic. A machine that society fervently desires cannot be ordered like a pizza. Edison spent years trying to invent a lightweight battery for electric automobiles that could be recharged quickly and could hold a charge for a long time. He made some progress, but 100 years later the problem still eludes complete solution.[18] Money and talent can speed refinements along, but they cannot always call an invention into being.

The internalist approach also emphasizes alternative solutions to problems. For example, late in the nineteenth century the need for flexible power transmission over a distance was solved by a variety of devices. In different places one could buy power in the form of compressed air, pressurized water, moving cables, steam, and electricity. These were not merely invented: by 1880 all were in commercial use. In Paris, compressed air drove machines in some small businesses. It was easy to use and far less trouble than installing and maintaining a steam engine. In New York and other nearby cities, the hot-air engine enjoyed a brief vogue. In Boston, from 1880 until as late as 1962 many small businesses in a single block had steam power delivered to them by overhead driveshafts. In hilly San Francisco, cable cars were superior to streetcars driven by electric motors.[19] The research of internalist historians explains the precise characteristics of these power systems, helps us to understand their relative merits, and shows us how they were used and why they were eventually abandoned.

The internalist can compare the technical merits of early steam-powered, electric-powered, and gasoline-powered automobiles, all of which were successfully produced and marketed between 1900 and 1920. This is necessary (but not sufficient) knowledge to understand why Americans preferred gasoline automobiles, even though initially they were more familiar with the steam engine. Only in retrospect was the gasoline engine the obvious winner. In 1900 the majority of the roughly 8,000 automobiles in the United States were steam-powered; gasoline autos were the least common type.[20] The electric car was the quietest and least polluting of the three, but it lacked the range of the others. Steam automobiles took the longest to start, as it took time to get cold water up to a boil. The Stanley Steamer overcame this objection by heating water in small amounts as it was needed. The steam auto was reliable, and people understood its technology.

Internal-combustion engines were noisy and polluting. Furthermore, there were few filling stations or mechanics to service them in 1900, while steamers burned kerosene, which was available in any hardware store. In addition, the early gasoline autos had to be cranked, which was inconvenient, physically demanding, and somewhat dangerous—the crank could kick back and hurt one's wrist or arm. On the other hand, because the steam auto was the heaviest of the three types, it was hard on driveways and road surfaces, and, at a time when most roads were unpaved, it easily got stuck in the mud. The batteries for electric cars were heavy and took a long time to recharge, making long trips inconvenient. The internal-combustion engine delivered the most power for its weight because its fuel had a high energy density. The internalist approach thus can identify the strengths and weaknesses of competing technologies. Clearly, the gasoline auto's longer range, lighter construction, and greater power gave it decided advantages. Yet such factors alone did not give it what Hughes would call technological momentum. By c. 1905 all three forms of automobile had been invented, had begun to spread into the society, and had experienced growth in demand. But none had gained a decisive lead, and all still competed with streetcars and bicycles.

To tell the story beyond this point requires the contextualist approach, which focuses on how the larger society shapes and chooses machines.[21] It is impossible to separate technical and cultural factors when accounting for which technology wins the largest market share. The Stanley brothers made fine steam cars, but, like most automakers, they built them by hand and in limited numbers. This meant that they were priced high. Electrics were manufactured in much the same way, so they had no price advantage. Some electrics were marketed as ideal for women, however. A sales strategy that "feminized" their ease in starting, lower speed, and limited range made such cars unappealing to men without attracting many female buyers.[22] Only the gasoline auto had an entrepreneur of the caliber of Henry Ford, who realized that the way forward was mass production of a standard design at the lowest possible price. As his managers invented and installed the assembly line, they brought the price of a new car down from more than $850 in 1908 to $360 in 1916.[23] Ford also benefited from geographical factors. The gasoline auto was best suited to use in the countryside, where the heavier steam cars sank into the mud and electrics could seldom be recharged. (Only one farmer in 15 had electricity.) In 1910 more than half the population of the United States remained on the land, and the gasoline auto had that market virtually to itself. Furthermore, rural people could better afford the Model T than its hand-assembled and more expensive competition. By 1926 an astonishing 90 percent of the farmers in Iowa owned automobiles,[24] and Ford had sold almost 15 million Model Ts. Ford's success spurred subsidiary investment in service stations, the training of thousands of mechanics, and the creation of a national network of companies selling tires, batteries, spare parts and the automobiles themselves. By the end of the 1920s, an auto industry overwhelmingly devoted to the internal-combustion engine consumed 20 percent of the nation's steel, 80 percent of its rubber, and 75 percent of its plate glass. Embedded in this extensive socio-technical system, the gasoline auto had achieved a technological momentum in the United States that it would not attain in Europe until the 1950s. In 1926, 78 percent of the world's automobiles were in North America. There was one for every six Americans, but only one for every 102 Germans.[25]

The success of the gasoline automobile can thus be attributed to a variety of interlinked factors. The lack of an electrical grid in large parts of the country and the unresolved problem of the heavy, slow-charging battery counted against the electric car, as did its extensive marketing as a woman's vehicle. The steam car had none of these problems, and the steam engine was familiar. Yet the steam car was the heaviest, it was not manufactured as cheaply or marketed as aggressively as the gasoline car, and gasoline was abundant and inexpensive. Thus, a wide range of factors were involved, including economics, entrepreneurship, and social norms as well as technology. In thinking about why AT&T's picture phone failed in the 1970s, Kenneth Lepartito concluded that any technology should be understood not as an isolated thing in itself, but as part of a complex system "in which machines have ramifications for other machines, for the plans of contending actors, and for politics and culture." Therefore, "all technological change becomes problematic. This indeterminacy flows from the fact that technology is not a stable artifact but a system in evolution, one whose features and functions are up for grabs."[26] But if a technology is widely adopted, this indeterminacy gives way to momentum.

In the contextual approach, every technology is deeply embedded in a continual (re)construction of the world. A contextualist eschews the Olympian perspective and tries to understand technologies from the point of view of those who encountered them in a particular time and place. This approach immediately implies that machines and technical processes are parts of cultural practices that may develop in more than one way. For example, the contextualist sees the computer not as an implacable force moving through history (an externalist argument), but as a system of machines and social practices that varies from one time period to another and from one culture to another. In the United States, the computer was not a "thing" that came from outside society and had an "impact"; rather, it was shaped by its social context. In this perspective, each technology is an extension of human lives.

The same generalizations apply to the Internet. Civilians under contract with the Department of Defense developed the Internet to facilitate communication among scientists using the large computers located at universities around the United States. The military funded it and understood its possible use in transmitting vital defense information in case of atomic attack. But the first working system connected universities, and when it was put into operation there was not a great deal of traffic.[27] No one had anticipated the most popular application: what we now call e-mail. In its early years, the system was funded by the Advanced Research Project Agency, out of the Pentagon, and was called ARPANET. In the 1970s, once it was up and running, the military tried to sell the system to AT&T, but AT&T refused the offer. For the next 15 years, scientists and grassroots organizations developed e-mail, user groups, and databases on the net.[28] In the 1990s came the World Wide Web, web browsers, and e-commerce. Then in a great rush the Internet became an integral part of advertising, marketing, politics, news, and entertainment. People used the Internet in unexpected and sometimes illegal ways. For some, "surfing" became a kind of tourism and an entertainment that partly replaced television. For others, the Internet offered ways to share music (often pirated), or to publish their thoughts and ideas in "blogs" (short for "weblogs") addressed to the world. The popular acceptance of the Internet raised political issues. Who should own and control it? Did it threaten to destroy jobs by eliminating middlemen, or was it the basis of a new prosperity? Did it democratize access to information, or did it create a "digital divide" between those who could afford it and those who could not? Like every technology, the Internet implied new businesses, opened new social agendas, and raised political questions. It was not a thing in isolation.

If one takes this approach, then it appears fundamentally mistaken to think of "the home" or "the factory" or "the city" as a passive, solid object that undergoes an involuntary transformation when a new technology appears. Rather, every institution is a social space that incorporates or doesn't incorporate the Internet at a certain historical juncture as part of its ongoing development. The Internet offers a series of choices based only partly on technical considerations. Its meaning must be looked for in the many contexts in which people decide how to use it. For example, in the 1990s many chose to buy books, videos, and CDs on the Internet, but not all were ready for the online purchase of groceries. By 2004 British supermarkets had wooed many customers to shop online for food, but the same idea had little success in Denmark, even though a higher percentage of Danes were online. People in many countries preferred the Internet to the fax machine, but use of the telephone continued to expand. People adapted the Internet to a wide range of social, political, economic, and aesthetic contexts, weaving it into the fabric of experience. It facilitated social transformations, but different societies incorporated it into the structures of daily life in somewhat different ways. Every culture continues to make choices about what to do with this new technology.

The history of electrification offers a suggestive parallel to the Internet. Between c. 1880 and 1920, when the electrical system was being built into American society, a lively debate took place among engineers, progressive reformers, businessmen, and the general public. This debate about the cultural meaning and uses of electricity was necessary because Americans of c. 1900 had to choose whether to construct many small generating stations or a centralized system, whether to adopt alternating or direct current, whether to rely on public or private ownership of the system, and whether to give control primarily to technicians, to capitalists, or to politicians. Similarly, the debate about the Internet (after c. 1992) was necessary because people had to decide whether to construct a decentralized or a centralized system. They debated to what extent it should be monitored or controlled by individuals, corporations, or the government, and they questioned to what degree the Internet and its many individual sites and homepages should become commercialized. Finally, they wrestled with the issue of proprietary software, shareware, and freeware, and with related issues of copyright and intellectual property.

Like the collaborative building of the Internet, electrification was not a "natural process" but a social construction that varied from country to country. For example, electrical light and power were embraced readily by the working class in Sweden, where the first educational publication of the Social Democrats (1897) was titled *Mere Ljus* (meaning "more light").[29] In contrast, South African mining towns used electricity almost exclusively to improve the extraction of gold, diamonds, and coal, and seldom to enhance the lives of black workers.[30] Yet not all consumers immediately embraced electricity. The majority of British workers long retained a familiar gas system. As late as 1936, only one-third of the dwellings in the industrial city of Manchester had electricity.[31] In the United States electrification proceeded much more rapidly, and more than 90 percent of urban homes had electricity in 1936.[32] Just as in the 1990s a divide opened between those who had computers and Internet connections and those who did not, during the 1930s there were equally worrisome divisions between those who had electricity and those who did not. The construction of the Internet opened up political issues, legal problems, entertainment possibilities, and business opportunities. As with electricity, the public found the Internet to be suddenly ubiquitous and yet inscrutable, and it often seemed to be an irresistible natural force. And as with electricity in 1910, so much was attributed to the Internet in the 1990s that it became a universal icon of the age.

The histories of the bicycle, the automobile, electrification, and the Internet all suggest, once again, that there is little basis for a belief in technological determinism. Sweeping externalist histories about machines that shape society remain popular, but they clash with the research of most professional historians of technology (both internalists and contextualists). The more one knows about a particular device, the less inevitable it seems. Yet a thousand habits of thought, repeated in the press and in popular speech, encourage us in the delusion that technology has a will of its own and shapes us to its ends. As we become accustomed to new things, they are woven into the fabric of daily life. Gradually, every new technology seems to become "natural," and therefore somehow "inevitable" because it is hard to imagine a world without it. Through most of history flush toilets did not exist, but after 100 years of widespread use they seem normal and natural; the once-familiar outhouse now seems disgusting and unacceptable. Likewise, Western societies have naturalized the radio, the mobile phone, and the television, and most people do not think of them as social constructions.

The novelist Richard Powers noted that the naturalization of technology involves the continual transformation of human desires "until in a short time consumers cannot do without a good that did not exist a few years before." As a contrast, Powers describes a character who has escaped from this continual process of naturalizing the new. She does not transform her apartment to accord with changing fashions, but keeps whatever she likes, regardless or style or age: ". . . Mrs. Shrenck's thing-hoard implied that she had bypassed this assimilation altogether, simply by making no distinction in value between a pine-cone picked up on yesterday's walk and a rare, ancient floor-cabinet radio. . . ."[33] Few consumers are so indifferent to style, however. And once one sees that technologies are shaped by consumption, it becomes apparent that, though devices change often, they do not necessarily improve. One of mankind's oldest technologies, clothing, offers a fine example, for stylistic considerations have often proved more important than comfort or durability. Judith McGaw examined the enormous variety of brassieres and found not only that there is "no convincing evidence that the breasts need support" but also that brassieres never quite fit. Women do not come in standard sizes, and furthermore "the size and shape of any woman's breasts change continuously—as she ages, as she gains or loses weight, as she goes through pregnancies, as she experiences her monthly hormonal cycles."[34] In selecting bras and other clothing, women continuously compromise between style, comfort, and self-expression. As this example suggests, the human relationship to technology is not a matter of determinism; it is unavoidably bound up with consumption.

In the excitement of the early Internet boom of the 1990s, a group of cyber-libertarians released a manifesto titled "Cyberspace and the American Dream: A Magna Carta for the Knowledge Age." Among their many claims was this assertion: "Turning the economics of mass-production inside out, new information technologies are driving the financial costs of diversity—both product and personal—down toward zero, 'demassifying' our institutions and our culture. Accelerating demassification creates the potential for vastly increased human freedom."[1] For decades, critics of industrialization had asserted just the opposite, arguing that mass production of goods and improved communications together annihilated cultural differences. Are advanced technologies being used to homogenize the world, dissolving distinctive cultures into a global system? Or are advanced technologies being used to create more social differences?

The preceding chapters established that technologies are neither deterministic nor predictable and that they are best understood as social constructions. These conclusions suggest that machines might be used to create choices and possibilities. A highly technological culture might become more diverse, and not, as many social scientists long assumed, more homogeneous. For more than 100 years, sociologists argued that industrial technologies were homogenizing people, places, and products. As the assembly line produced identical goods, it seemed to erase difference. Workers became interchangeable, and consumers with identical houses and cars seemed interchangeable as well. As technical systems became more complex and interlinked, the argument ran, people became dependent on the machine and had to adjust to its demands. Technology shaped the personality and dominated mental habits. Veblen declared: "The machine pervades the modern life and dominates it in a mechanical sense. Its dominance is seen in the enforcement of precise mechanical measurements and adjustments and the reduction of all manner of things, purposes and acts, necessities, conveniences, and amenities of life, to standard units."[2] For Veblen's generation, the assembly line and mass production became powerful metaphors for standardization and machine domination.

The homogenizing effects of technology seemed most obvious in the United States. In *Life and Thought in America,* Veblen's contemporary Johan Huizinga wrote: "The progress of technology compels the economic process to move toward concentration and general uniformity at an ever faster tempo. The more human inventiveness and exact science become locked into the organization of business, the more the active man, as the embodiment of an enterprise and its master, seems to disappear." Huizinga noted the attraction of interchangeability and argued, like Alexis de Tocqueville before him, that "the American *wants* to be like his neighbor." Indeed, he wrote, the American "only feels spiritually safe in what has been standardized."[3] By 1918 the uniformity had become so great that a visitor traveling a thousand miles from one city to another inside the United States often was disappointed because one city seemed so much like the other. In the 1930s and later, the Frankfurt School viewed modern communications as the primary means of creating uniformity and social control.[4] They feared that as publishing, radio, and film penetrated popular culture they packaged, standardized, and trivialized human complexities. Max Horkheimer complained to a colleague: "You will remember those terrible scenes in the movies when some years of a hero's life are pictured in a series of shots which take about one minute or two, just to show how he grew up or old, how a war started and passed by, and so on. This trimming of existence into some futile moments which can be characterized schematically symbolizes the dissolution of humanity into elements of administration."[5] From such a perspective, cultural power had passed to the dream factories of Hollywood and the songsmiths of Tin Pan Alley.

The modern man portrayed in David Riesman's book *The Lonely Crowd* (1950) no longer actively directed his life but was shaped by forces and movements outside himself.[6] It seemed obvious that the more technological a society became, the more uniform was its cultural life. "As for pluralism," George Grant complained in 1969, "differences in the technological state are able to exist only in private activities: how we eat; how we mate; how we practice ceremonies. Some like pizza; some like steaks; some like girls; some like boys; some like synagogue; some like the mass. But we all do it in churches, motels, restaurants indistinguishable from the Atlantic to the Pacific."[7]

In *Culture against Man,* a widely discussed book published in 1963, Jules Henry argued that modern people had acquired destructive "technological drives." These "drives," he asserted, "can become almost like cannibals hidden in a man's head or viscera, devouring him from inside. . . . The American then may consume others by compelling them to yield to his drivenness." These technological values were also self-destructive, in Henry's view: "Americans get heart attacks, ulcers, and asthma from the effects of their drives, and it seems that as exotic cultures enter the industrial era and acquire drive, their members become more and more subject to these diseases."[8] Henry argued that science and technology had become the center of a "culture of death."

The student revolt of the 1960s was in part a rejection of the values of efficiency and standardization. Mario Savio attacked the University of California for conceiving itself as a factory in which the president was the manager, the faculty were the workers, and the students were raw materials to be manufactured into docile parts of "the machine." "There is a time," Savio declared to cheering Berkeley students, "when the operation of the machine becomes so odious, makes you so sick at heart, that you can't take part; you can't even passively take part, and you've got to put your bodies upon the gears and upon the wheels, upon the levers, upon all the apparatus, and you've got to make it stop. And you've got to indicate to the people who run it, to the people who own it, that unless you're free, the machine will be prevented from working at all!"[9]

Theodore Roszak's 1969 book *The Making of a Counter Culture* also attacked the attempts of government and corporate experts to manage and standardize daily life. Roszak declared that the "prime strategy of technocracy" was "to level down to a standard of so-called living that technical expertise can cope with—and then, on that false and exclusive basis, to claim an intimidating omnicompetence."[10] A popular song from the same period, written by Malvina Reynolds, complained of towns of "little boxes" and described how the identical people living in these boxes had identical children who all went to the university, where they too were put into "little boxes, all the same." The 1960s student revolt was far more than a reaction against institutionalized racism or against the Vietnam War. It was also a reaction against standardization, efficiency, and business-directed routines that young radicals felt had come to dominate their lives.

One community seemed to epitomize the world of "little boxes": Levittown, New York, built after World War II.[11] The same builders constructed all the houses at the same time, using mass-production techniques that they had developed working under government contracts during the war. In several locations, Levitt's company eventually would put up more than 140,000 houses, using identical parts that were cut or fabricated before being delivered to the construction site. Levittown, New York, with 17,400 houses, was the largest development yet put up by a single company. Buyers literally lined up to purchase the homes, in part because they got more floor space for their money than elsewhere. Yet many criticized the development. Wouldn't such uniformity produce standardized, soulless people? Riesman compared suburban life to a small college fraternity.[12] And Levittown was not an isolated example. Similar planned communities were built in much of the Western world. Standardization seemed to be running rampant.

Today Levittown is not a monotonous row of "little boxes." Homeowners have added garages, pillars, dormers, fences, and extensions. They have painted their homes many different colors, and planted quite different shrubs around them, landscaping each plot into individuality.[13] In 2006, a visitor to Levittown has to study the houses carefully to see their common elements. Three generations of homeowners have used a wide range of technologies to obliterate uniformity. Starting out in the 1950s with a relatively homogenous population that was white, middle-class, and young, the mass-produced suburb of Levittown has become increasingly diverse in appearance, demographics, and racial makeup. A house that sold for $6,700 in 1952 was worth $300,000 half a century later, and it could be sold immediately.[14]

The reshaping of Levittown has equivalents elsewhere, notably in the automobile industry. During the first years of the assembly line, Henry Ford refused to manufacture a wide variety of cars. Instead, the Model T was available in only a few variant forms and a limited range of colors. Ford was reputed to have said that customers could get any color they wanted, so long as it was black. In fact, a few other colors were available in the 1920s. Nor was the model design completely static, as the company constantly made small improvements. But Ford eschewed the annual model change, because it was expensive to retool the assembly line to accommodate changes in a car's external appearance. By freezing the basic design, Ford could concentrate on improving the efficiency of his assembly line, which in turn made it possible to lower the cost of his cars. A new Model T dropped dramatically in price, from $850 in 1908 to $360 in 1916, with further reductions into the 1920s.[15] In contrast, General Motors brought out new models every year, in a changing variety of colors. To do this, GM annually retooled its assembly line, passing the cost on to the consumer. To Ford's dismay, the public embraced annual changes, and gradually GM won so much of the market that in 1927 Ford reluctantly abandoned the Model T (after producing more than 15 million) and began to make annual models with greater variety.[16] Today, even the least expensive cars come in many colors, and mid-range automobiles can literally be made in hundreds of different ways, depending upon the upholstery, accessories, colors, and options a buyer chooses. In 2004, the Ford F-150 pickup truck was available in 78 different configurations that included variations in the cab, the bed, the engine, the drive train, and the trim as well as in the colors of the upholstery and the exterior paint.[17] And once a vehicle was purchased, the owner could customize it further to the point that it literally was one of a kind. The telephone provides another example of differentiation replacing standardization.[18] Early in the twentieth century, AT&T completely dominated telephone design, and virtually all Americans had black desktop telephones. Engineers had designed them to be long lasting and functional, and the consumer had no choice. Indeed, the consumer did not even own the home telephone, but rented it by the month. Engineering concerns for sound quality and national service predominated over questions of design. AT&T sought profits not by selling a product that soon became obsolescent, but by renting a product that was extremely durable. It developed not a diversity of telephones but the world's most extensive system, serving a large customer base. However, just as Henry Ford could not keep selling only Model Ts, eventually "Ma Bell" had to pay more attention to consumers. For several generations the major innovations in the telephone were technical: better sound quality, direct dialing, better long-distance connections, and so forth. In contrast, by the 1950s Europeans had telephones with more visual appeal, though the service itself almost always cost more. In the 1960s AT&T also began to offer a wider variety of telephone colors and styles. It encouraged families to acquire different designs for kitchens, living rooms, and bedrooms. This successful campaign was only the beginning. Once consumers could own telephones rather than rent them, thousands of different designs appeared. Novelty phones were sold that resembled almost any conceivable object. In the 1980s, when AT&T was split up, service companies proliferated.

At the same time, telephones went from being collective to being individual. In 1950, it was still common for several families to share a party line. By 1970, most households had a private line, shared only by family members. In subsequent decades, separate phones for different members of the family, especially teenagers, became more common. But the full differentiation of the telephone came with the proliferation of mobile phones, each owned by one person who carried it all the time. Not only did many firms produce mobile phones; they constantly upgraded the models so they seemed to have as many options as a sports car. Not only did consumers have distinctive phones; they continually changed their ring tones, which by 2004 had become a major source of revenue for the music industry. In short, as with Levittown and the Ford, the telephone metamorphosed from a uniform mass-produced item into a highly differentiated product.

No institution better understood consumer demand for variety than the department store. Department stores emerged in the second half of the nineteenth century, notably in France, Britain, and the United States but also in Germany, Scandinavia, Canada, and elsewhere. A department store was a vast commercial space that dwarfed the traditional shop. Department stores competed not only on price but also on selection, giving consumers choices among products differentiated into the widest possible range of styles and qualities. Managers found that consumers' tastes changed rapidly, and it became their business to shift their assortment accordingly. By the early years of the twentieth century, manufacturers routinely relied on department stores' managers and buyers to find out what the public wanted.

"Fashion intermediaries" became crucial interpreters of consumers' desires. As Regina Blaszczyk concluded, "supply did not create demand in home furnishings, but demand determined supply."[19] Blaszczyk's book *Imagining Consumers* shows how firms catered to the mass market through an interactive process of discovering what consumers wanted and then delivering it quickly through flexible production. Department stores had to stock more than mass-produced goods, because their customers did not want to dress or furnish their homes all in the same way. Consumers' demands were relayed to factory designers by salesmen, retail buyers, materials suppliers, art directors, showroom managers, home economists, advertising executives, and market researchers. The "field letters of factory salesmen read like primitive market research reports."[20] A wholesale jobber in Denver listened to country storekeepers as they inspected wares and then reported their comments back to the factory. The department store did not dictate style to the consumer so much as it relayed changes in taste to the manufacturer. As an executive in the advertising firm J. Walter Thompson put it in 1931, "the consuming public imposes its will on the business enterprise."[21]

Because consumers demanded differentiation, mass production was not the only way or necessarily the most profitable way to manufacture. Although early-twentieth-century critics were mesmerized by assembly lines and giant corporations such as Standard Oil, Ford, and General Motors, Philip Scranton (one of the leading scholars in the history of technology) has shown the importance of mid-size and smaller firms that made a wide variety of consumer goods through customized and small-batch production. Far from being backwaters destined to be rationalized by scientific management or converted to assembly lines, such companies were just as important to an advanced economy as the large corporation. Because they were smaller than Ford or GM, and because they tended to remain family firms or closely held corporations, they have not attracted so much attention. But collectively, in 1909, such firms contributed one-third of the value added in the US economy as a whole, and employed one-third of all workers.[22] Their specialty production grew just as fast as mass production, and they added more value and employed more workers. By 1923 value added had tripled and specialty firms still accounted for one-third of industrial employment. They did not produce identical goods, but responded flexibly to demands for variety. Such companies were innovative and profitable, and they made possible the endless novelty that was the hallmark of a consumer society. With little use for standardization, they explored many different production systems. They might use scientific management or other systematic approaches to increase reliability, to reduce errors, or to reorganize, but they resisted changes in the organization of work that froze design or constrained innovation.[23] These companies excelled in the batch production of such items as carpets, furniture, jewelry, cutlery, hats, and ready-to-wear clothing. Then as now, middle-class consumers wanted new styles, and large profits accrued to the firms with flexible modes of production that could supply differentiated product lines to retailers.

The social critics of the first half of the twentieth century overstated the degree to which mass production would be accompanied by conformity and standardization of the personality. Neo-Marxists were particularly likely to overstate the case.[24] In fact, consumers relentlessly demanded variety, and over time, even the companies and products that seemed to epitomize mass society dissolved into difference. Ford had to abandon the Model T, the universal black telephone from AT&T evolved into a myriad of sizes, colors, and designs, and Levittown's uniformity disappeared in a wave of home improvements and landscaping.

The shift from viewing technology as the harbinger of standardization to viewing it as the engine of differentiation is epitomized by attitudes toward the computer. In the 1950s, the computer represented centralized control, systematic information gathering, and the invasion of privacy. It apparently promised the ultimate negation of the individual, making possible an extension of the rationality of the assembly line and its interchangeable parts into new areas. The first pages of Francois Lyotard's book *The Postmodern Condition* (1984) address the computer's effects on society and the "hegemony of computers." According to Lyotard, computers bring with them "a certain logic, and therefore a set of prescriptions determining which statements are accepted as 'knowledge' statements."[25] Lyotard portrays the computer as an apparatus of top-down control in the creation and maintenance of a new circulation of information likely to benefit large institutions.

After c. 1980, however, the computer, once feared as the physical embodiment of rationalization and standardization, gradually came to be seen as an engine of diversity. The decisive move was from giant mainframe computers that stored all the data and software at a few central locations to the personal computer, which decentralized the system. As individuals gained control over their own machines, they could personalize them by choosing their favorite font, downloading their favorite songs, or using works of art as screen savers. Gradually, the computer ceased to seem threatening or external and became a convenience at work and a partner in all sorts of play. In the 1990s, millions of individuals began to enjoy the possibilities of e-mail and the World Wide Web. Each person could, and most students did, construct a homepage and make it available to the world.

However, not all groups have equal access. To what extent is the Internet limited to affluent and well-educated consumers? Is there a digital divide? In the United States in the year 2000, 56.8 percent of Asian households had Internet access, compared to

29.3 percent for African-Americans and only 23.7 percent for His-panics.[26] Income accounts for some of these differences, but the cultural values of minority groups are also important. Among households with incomes below $15,000, one-third of the Asian-American households had Internet access; only one-sixteenth of the African-American households and one-nineteenth of the Hispanic-American households had it. Given the same resources, these groups have different priorities. The gap remained much the same at all income levels, with Asian-Americans and whites on the one side of the digital divide and with African-Americans and Hispanic-Americans on the other.[27] (The digital divide was not gendered; 44 percent of both men and women had Internet access.[28])

Internationally, the differences appear to be far greater. In 1999 half of the world's population had never made a telephone call, and less than 1 percent of India's population was on the Internet.[29] Rapid diffusion of technologies to all people everywhere can hardly be taken for granted. Rather, for most of the world, use of the Internet seems limited to the educated and wealthy in urban areas. Yet matters are not so simple. At the end of the 1990s two anthropologists, Daniel Miller and Don Slater, studied the Internet on the island of Trinidad, in order to test the widespread assumptions about its unequal distribution and social effects.[30] Their work suggested that the Internet is more widely diffused than one might expect, but that it is not erasing cultural differences. They looked to see if Internet use was strongly marked by differences in wealth, since computers, modems, and time online are expensive. They found that Internet use in Trinidad was quite democratically dispersed through Internet cafés. They had expected that businesses (notably advertising agencies and telecommunications companies) might be driving forces in Internet development. They were not. On Trinidad, the Internet was far less a project of capitalist hegemony than a grassroots movement. Network use did not conform to expectations of globalization, either. Users did not create new online identities that were more international than their offline sense of self. Rather, they used the Internet to maintain and strengthen a web of relationships with extended families in Britain, Canada, and the United States.

Nor did the Trinidadians embrace a global culture that weakened their sense of identification with their own nation or culture. Indeed, it was difficult to find islanders who feared that the Internet might absorb them into a global mass culture. On the contrary, they used the Internet to project pride in their own nation, to broadcast its music, to educate others about their islands, and to sell its products and vacation experiences. The Internet did not seem to overwhelm but rather to strengthen the local, for example making it easier for the music of Trinidad and Tobago to reach the rest of the world. Miller and Slater found that using the Internet did not even seem to undermine the local dialect, as the "Trini" idiom flourished on personal homepages. In short, use of the Internet strengthened and spread local cultural values rather than undermining them. The uses of the Internet on Trinidad suggest once again that people adapt technologies to express local identities.

The technological specifications of the Internet have to be much the same in Trinidad as elsewhere for the system as a whole to work, but the social construction of the Internet can vary a great deal. Likewise, the design and the mechanical functioning of the automobile are the same in Denmark as in the United States. However, the place of the automobile in the two societies is not the same. In Denmark, it is quite possible for a man or a woman to rise to the highest positions of power and influence, such as the president of a university or head of a government ministry, without learning to drive an automobile. This would be exceedingly unlikely in the United States, where the automobile is a necessity in order to transact fundamental tasks of everyday life.

Recently, debates about globalization have addressed the question of whether technologies are being used to create a more homogeneous or a more heterogeneous world. George Ritzer has warned of "McDonaldization," arguing that fast-food restaurants epitomize an impersonal standardization that Western nations aggressively export to the rest of the world.[31] Similarly, Benjamin Barber has attacked the cultural imperialism that, he argues, is transforming a rich cultural variety into a single bland "McWorld."[32] Francis Fukuyama disagrees with Ritzer and Barber. He does not see cultural imperialism; he applauds the triumph of free markets, democracy, and Western cultural values.[33] All these authors are externalists with little interest in technology per se.

Other critics, such as Roland Robertson, find that large corporations and Western-style corporations do not obliterate local cultural constellations.[34] Instead, each local culture adopts only some of the products and practices on offer in the global marketplace, adapting those it has selected to fit into its own routines. Rather than adjust to a single pattern, each cultural region creates hybrid forms, which Robertson calls "glocalization." Even McDonald's finds that it must give in to this process. In Spain it sells red wine to go with its hamburgers, and in India (where cows are sacred) it does not serve beef. More generally, an endless process of creolization is taking place, producing such novel combinations as Cuban-Chinese cuisine, Norwegian country music, and "Trini" homepages. As Rob Kroes imaginatively argues, in cultural contact zones, people "are scavenging along the tide line of Western expansion, appropriating its flotsam and jetsam. They feel free to rearrange the order and meanings of what they collect. They turn things upside down, beads turn into coinage, mirrors into ornaments. Syntax, semantics, and grammar become jumbled." In this process of selective appropriation, "people at the periphery create their own environment" and this creolized cultural production often may be re-exported.[35] Just as in the United States former slaves and immigrants created their own cultural worlds, selectively appropriating elements of different cultures, so too cultures that come into contact with Western society engage in a creolizing process. What results is not a standardized world but a potentially endless process of differentiation.

Just as mass production did not obliterate batch production, which was more flexible than might have been expected and which responded to changing consumer demands, fears of standardization were exaggerated. "Glocalization" or creolization is common. Though critics may focus on the supposed homogeneity of a machine culture, the owner of every house, car, and telephone in Levittown can express individuality through consumption, and every cultural group can do likewise. Multiculturalism was not founded on a rejection of technology; rather, it emerged precisely at the time when advanced methods of production and distribution made it easier to manifest difference. The same change can be seen in television's transition from a small number of dominant broadcasters to "narrowcasting." In the 1950s, an oligopoly of three national networks dominated American television, and many European countries had only state television. Half a century later, the average home receives dozens of channels, including stations specializing in history, science fiction, old films, sports, nature, foreign language programs, and many other areas. The same process has emerged in marketing, where extensive segmentation divides consumers into smaller target groups rather than seeking to sell one thing to everyone. Whereas the early twentieth century was the era of mass production for a mass market, in recent decades advertisers have become more discriminating. Rather than send out the same brochure to millions of homes, they compile specialized mailing lists, they divide the market by postal codes, or they use niche cable TV stations that address only a fraction of the national market. Market segmentation reached its logical culmination with the Internet, where the more sophisticated homepages (such as Amazon's) interact with customers and suggest new products to them on the basis of their previous purchases.

Resistance to standardization also means that much shopping continues in real space and time, because consumers want to see and touch before they buy. To satisfy the heterogeneous demands of consumers, entrepreneurs have erected larger and larger venues. After World War II, they built supermarkets and shopping centers that replaced neighborhood stores. In the 1980s, developers began to build huge malls in order to provide the widest possible variety of shops and services. The arrays of goods encourage each person to construct or invent an identity of choice.

The expansion of the supermarket also registers the emergence of multiculturalism. The production and consumption of food has a central place in every culture. As the world's most energy-intensive society, the United States uses 17 percent of its power for food. Roughly one-third of this is used for food production, one-third for manufacturing and processing, and one-third for transportation, refrigeration, cooking, and dishwashing. Ethnic differentiation flourishes within this energy-intensive framework. By the 1980s, a typical large supermarket stocked 30,000 items. The supermarket made possible millions of possible combinations of items in multicultural or hybrid meals reflecting several backgrounds in a single household. The price of this variety was a decline in the number of stores, while the average floor space of a single store doubled. At the supermarket, multiculturalism and advanced technological systems seem compatible. This argument suggests that ethnic, racial, and regional identities are triumphing over national uniformity, and that consumers' demands are making advanced technologies into tools for heterogeneity instead of standardization.

There is a counterargument. The ethnic variety on the supermarket shelves is based on capitalist rationalization, packaging, and distribution. The system as a whole includes not only the products and the shoppers but also investors seeking maximum profits, the agricultural sector, the food-processing industry, and the trucking industry. To see why this matters, consider the visitors to the immigration museum at New York's Ellis Island. While there, they can have lunch in a food court much like that in any large shopping mall. On one level, the variety of food seems to exemplify the multiculturalism that emerged from immigration. The Ellis Island food court offers choices between ethnic cuisines, and quite literally does not represent the melting pot. However, on the level of the technological systems used to produce and deliver the food, most differences evaporate. The food court's businesses all use the same kinds of freezers, steam trays, fryers, and microwave ovens. They prepare dishes suited to the demands of a cafeteria, an assembly-line operation that functions best when food does not require much on-site preparation before it is served. They all serve food on disposable plates. The foods may not taste the same, but some of the most pungent differences have been toned down or eliminated to reach a wider customer base. The customers eat these meals quickly and in the informal American manner. If one focuses on process rather than content, these meals are produced, served, and consumed in much the same way.

Though every individual can select from a wide range of foods or cars or telephones, each is still enveloped in larger technological systems. It is easier to select among many telephones than it is to do without one. It is easier to make choices at the supermarket than to find an alternative source of food. And it has become almost impossible in much of Western society to live without an automobile. Such patterns have larger implications. Whatever food or automobile they prefer, Americans of all ethnic and racial backgrounds get less exercise than they once did, and they tend to eat high-sugar and high-fat foods that lead to obesity and the "diseases of affluence."

Furthermore, the majority of ethnic restaurants depend on customers from outside the group that the cuisine "represents." Indeed, "55 percent of America's consumer food budget is spent on restaurant meals and ready-to-eat convenience foods."[36] This is a higher percentage than in most other countries, and it is a further indication that market forces underlie the display of ethnic diversity. In addition, the food is often modified to appeal to the taste of a wider public. For example, Mexican food served abroad is often not quite so fiery as that served in Mexico, and the Swedish smorgasbord in North America often only vaguely resembles what Swedes eat at home. The surface appearance of multicultural variety should not obscure the homogenizing practices involved in marketing, advertising, and cooking cuisines so they will appeal to the general public.

The cyber-libertarians were only partially correct when they predicted a "demassification" of production that would increase differentiation and provide greater scope for the construction of new identities. Consumer liberation to a considerable degree is re-contained, or limited, by the technological momentum of institutional structures. Within such structures, after c. 1980 consumers demanded more racial and ethnic differentiation in the marketplace. Indeed, the greater the pressures of globalization, the more attractive "difference" became.

Multiculturalism as a whole might be seen as part of the "invention of tradition" argument put forward by Eric Hobsbawm and Terence Ranger. Much of what seems the venerable survival of ancient customs turns out to have been shaped or even created wholesale by nineteenth-century nationalists intent on establishing a pedigree for a certain cultural group, for example through the promotion of national flags, the recovery and promotion of folk dances and traditional clothing, or the writing of national anthems. Hobsbawm and Ranger cite "the use of ancient materials to construct invented traditions of a novel type for quite novel purposes," noting that "a large store of such materials is accumulated in the past of any society" and that at times "new traditions could be readily grafted on old ones . . . by borrowing from the well-supplied warehouses of ritual, symbolism and moral exhortation—religion and princely pomp, folklore and freemasonry (itself an earlier invented tradition of great symbolic force)."[37]

Technologies are related to this process in several ways. First, it is often the case that technologies have been used to disrupt the social fabric and to undermine customs, creating a need for new, invented traditions as substitutes for lost routines and undermined social patterns. Indeed, Hobsbawm and Ranger argue that the invention of tradition will "occur more frequently when a rapid transformation of society weakens or destroys the social patterns for which 'old' traditions had been designed."[38] This has been particularly the case for the last 200 years. Second, newly invented traditions are almost always disseminated and discussed through the media. Third, if embracing new traditions can be enhanced through tourism, then the steamboat, the railroad, and later the automobile and the airplane were essential to mass participation. For example, in Britain certain nineteenth-century physical structures have become symbols of the nation, notably Big Ben, the Houses of Parliament, and Trafalgar Square. The patriotic British or Imperial subject who lived outside the capitol saw innumerable photographs of these sites, and wanted to see them in person. The national transportation network made it easier to satisfy this desire. Not incidentally, many of the railway stations, tunnels, and bridges on this journey also become images of national greatness.

Just as markers of Scottish nationalism such as elaborate varieties of tartans for every clan were invented in the nineteenth century, in the second half of the twentieth century multicultural groups searched their histories and invented traditions to underpin a separate sense of identity. Marketers were only too pleased to oblige new preferences, and advertisements of the 1980s and the 1990s stressed visible differences. Whether viewed as corporate co-optation (Naomi Klein) or as creative appropriation (Rob Kroes), a flourishing multiculturalism has become part of the invention of traditions that has characterized the West since industrialization. Instead of creating traditions that reinforced a unifying nationalism, however, the multicultural project invented and celebrated racial, ethnic, or regional diversity.

Such variety is extravagant. If technology does not automatically create crushing uniformity and standardization, does the environment set limits on how much difference we can express?

6 Sustainable Abundance, or Ecological Crisis?

Does mastery of technology ensure abundance? Daniel Defoe's *Robinson Crusoe* (1719) answered unequivocally that it does. Based on actual experiences of a castaway British seaman, the novel describes its hero's 26 years on a deserted island off the coast of South America. Starting with nothing but a gun and a few tools salvaged from the wreck of his ship, the fictional Crusoe creates a comfortable though lonely life. In the eighteenth century, his story seemed a parable about the superiority of Western civilization. Crusoe's metal tools and weapons alone do not make him superior. Far more important, he has inherited generations of technical experience. He knows how to build a fortified shelter, and he does not have to invent the idea of a table, a chair, or other furniture. He catches fish, tames goats, weaves baskets, makes pottery, and in the course of several growing seasons patiently converts a few seeds of wheat into an annual crop. *Robinson Crusoe* became popular just as Britain began to industrialize. In subsequent generations, the steam engine promised radical increases in power and productivity, leading a late-eighteenth-century poet to declare: "Ingenious engines wondrous works perform, The hungry nourish and the naked warm!"[1]

For liberal political thinkers, the difference between Crusoe and the uneducated native whom he trained as his servant measured the distance from primitive society to civilization.[2] From a liberal perspective, society may have had a rough equality in pre-history, but the extent of a society's possessions measures its advance. At the most primitive stage of existence, in other words, people are all equally impoverished. Progress requires the extraction of a surplus, and at times this has involved slavery or feudal bondage. Over millennia, however, society became more rational, as economic markets and laws, instead of naked power, became the basis for exchange. Technological liberals believe in humanity's ability to improve both production and distribution. Although social classes persist, they argue, the life of the average person continues to improve. The popular expression "a rising tide lifts all boats" expresses this optimism. The widespread liberal view has long been that advances in technology bring greater efficiency and prosperity for all in the form of higher wages, less expensive goods, better transportation, and shorter working hours. Most mainstream European and American politicians of the twentieth century embraced this view. Even Marxists expressed similar ideas, although they assumed that the full advantages of the machine would not be realized until a revolution had occurred. As late as 1970, Herbert Marcuse declared: "I believe that the potential liberating blessings of technology and industrialization will not even begin to be real and visible until capitalist industrialization and capitalist technology have been done away with."[3]

Visions of technological progress seem attractive in the abstract, but what do they mean in practice? The landscape provides a physical measure of technological change. One of the founders of landscape studies, J. B. Jackson, defines landscape as "a composition of man-made or man-modified spaces to serve as infrastructure or background for our collective existence."[4] Landscape is not natural; it is cultural. It is not static; it is part of an evolving set of economic and social relationships. Landscapes are part of the infrastructure of existence, and they are inseparable from the technologies that people have used to shape land and to shape their vision. People continually put the land to new uses, and what appears to be natural to one generation is often the product of a struggle during a previous generation. Some of the apparently wild moors beloved of hikers in Britain were once thickly forested. In other parts of England and Scotland, land-owners evicted small farmers during the period of enclosure, creating a countryside that visitors now take to be "natural." In Denmark, wheat-growing areas were converted to dairying during the nineteenth century because of pressure from inexpensive grain produced in the New World. The same pressure forced many New England farmers to give up agriculture. Most of the forested hillsides in New England were cleared pastures in 1840. Almost everywhere, the appearance of the land is the result of an inter-regional interplay between agriculture, industry, and leisure activities. Technologies also affect the air, which carries traces of smoke, microscopic particles, pollen, carbon monoxide, and the dust raised by travel. Even at national parks and in supposedly untouched wilderness areas, the intermingling of culture and nature continues, whether unintentionally (as in the cases of acid rain, pollution, and migration of plant species) or intentionally (trails, fire breaks, campsites, roads, scenic outlooks).

Landscapes express the technologies and land use of earlier generations. The farms in much of Europe and along the Atlantic coast of the United States are pastoral landscapes of irregularly shaped fields, pastures, and woodlots, each fitted into the form of the land. Their modest scale is attributable to their original use in family farming with draft animals. These pastoral landscapes proved reliably efficient for hundreds of years, though in many areas they have now been abandoned (as a result of competition from more industrialized agriculture elsewhere) and gradually reforested. Yet this pastoral landscape is still productive enough to sustain the Amish or the Mennonites with a minimum of modern technology.

The landscapes that out-competed rural New England—the flatter and more fertile Middle West, the Canadian prairies, the pampas of Argentina—were farmed with new kinds of agricultural machinery. In 1890 a single US farmer could produce 380 bushels of wheat in the same time his grandfather needed to produce 20 bushels in 1830.[5] The productivity gains were greater for barley, less for other crops, but in all cases at least 200 percent. The new machines substituted horsepower for manpower, and worked best on large, flat, square fields. Productivity has increased just as dramatically since 1890, owing to gasoline motors, tractors, electrification, new fertilizers, and hybrid seeds.

Irrigated farms in California and Arizona take the rationalization of landscape further. Modern agribusiness relies on electric pumps that water fields when sensors say it is time. Airplanes spray the fields to keep down weeds and kill insects, and a combination of high-tech machines and migrant labor harvests the crop. The fields are enormous and laid out with mathematical precision, both to ensure proper irrigation and drainage and to let machines move efficiently over them. Such farming is profitable as long as energy and irrigation water remain inexpensive.

This succession of landscapes, from pastoral New England to industrialized family farming in the Middle West to irrigated agribusiness, traces an arc of increasing productivity made possible by increasing investment in mechanization. In the first half of the nineteenth century, many believed such agricultural development illustrated how industrialization creates more wealth, more jobs, and more goods for all. In the United States, Daniel Webster and Edward Everett were particularly effective spokesmen for manufacturing. They argued that more technical skill and more mechanical power led to a higher level of civilization. By 1900 such views had become orthodox. Robert Thurston, a specialist in textile production who served as president of the American Society of Mechanical Engineers, quantified the argument.[6] Between 1870 and 1890 he saw factory productivity rise almost 30 percent, while working hours dropped and real wages rose 20 percent. Similar gains throughout the nineteenth century had radically improved daily life. People had more leisure time and more money to spend. Consumption of clothing, appliances, and home furnishings rose rapidly, and Thurston drew graphs to express "the trend of our modern progress in all material civilization." "Our mills, our factories, our workshops of every kind," he wrote, "are mainly engaged in supplying our people with the comforts and the luxuries of modern life, and in converting crudeness and barbarism into cultured civilization. Measured by this gauge, we are fifty percent more comfortable than in 1880, sixteen times as comfortable as were our parents in 1850, and our children, in 1900 to 1910, will have twice as many luxuries and live twice as easy and comfortable lives."[7]

Thurston exaggerated, but Americans had made considerable progress in acquiring material goods. When Werner Sombart visited America (preparing to write *Why Is There No Socialism in the United States?*), he observed that American workers had greater material well-being than their European counterparts.[8] Wages were two to three times higher than in Germany, yet food and clothing cost roughly the same.[9] American workers dressed better, ate better, and were more likely to own a home. They were far less likely to live in a tenement, and one- and two-family houses were widespread. Sombart did not argue that the greater material well-being of American workers by itself explained the failure of socialist movements to achieve as much political power as in Europe. Ethnic and religious divisions, a modicum of upward mobility, and a different political system were also important. Furthermore, adults of that era were deeply impressed by inventions such as the telephone, the motion picture, the phonograph, the x-ray machine, the automobile, and the airplane. These inventions did not remain remote. They were not limited to the wealthy, but became democratically dispersed. In 1911 the president of MIT wrote: "Our grandfathers, looking down upon us, would feel that they observed a new heaven and a new earth."[10] The pace of technological change was undeniably quickening, and ordinary people acquired phonographs, electric lighting, and automobiles. In the early decades of the twentieth century, with such palpable evidence of change, Americans made engineers into cultural heroes in popular novels and films.[11]

In 1926, Henry Ford, whose factories had pioneered the assembly line and built more than half of the world's automobiles, declared (with the help of a ghost writer) that the machine was "the symbol of man's mastery of the environment."[12] For Ford, human history was about the development of power, from the "laborious hand culture of the soil" that he knew firsthand from a rural childhood to industrial society. His best-selling book declared that the modern mastery of power "would increase and cheapen production so that all of us may have more of this world's goods." This hardly seemed an empty promise coming from a man whose Model T cars got cheaper each year while improving in quality. For Ford's generation, nature was no longer outside human control; it was a source of raw materials to be exploited for human development.

Between 1851 (the year of London's Crystal Palace Exposition) and 1958 (the year of the World's Fair in Brussels), technological optimism found expression in a series of world's fairs on both sides of the Atlantic. All embodied the belief in material progress based on technology. In model homes, cities, and farms, fairgoers glimpsed an improved future not as an abstraction, but materialized in prototypes and demonstrations. Americans experienced their first telephone at the Philadelphia Centennial Exposition in 1876, their first electrified cityscape at the Chicago Columbian Exposition of 1893, their first full-size assembly line and first transcontinental telephone call at the San Francisco Panama Pacific Exposition in 1915, and their first television at the New York "World of Tomorrow" Exposition of 1939.[13] Europeans also celebrated these technologies, and in vast imperial pavilions showed how essential they were to their spreading empires.[14] Chicago's "Century of Progress" Exposition (1933) marked the growth of a small village into the second-largest city in the United States. Its exhibits were planned to "show you how Man has come up from the caves of half a hundred thousand years ago, adapting himself to, being molded by, his environments, responding to each new thing discovered and developed. You see man's march upward to the present day. . . ."[15] The Chicago fair's promoters announced the pre-eminence of the machine in the slogan "Science Finds—Industry Applies—Man Conforms."

Technological optimism may have reached its peak in the middle of the twentieth century. The march of progress then seemed to lead to a work week of 30 hours or less, early retirement, and a life of leisure and comfort for all. In 1955 Congress was "told by union leaders that automation threatens mass unemployment and by business executives that it will bring unparalleled prosperity."[16] Most believed, however, that leisure could be properly managed, and 1968 testimony before a Senate subcommittee "indicated that by 1985 people could be working just 22 hours a week, or 27 weeks a year, or they could retire at 38."[17] Some social scientists proclaimed that control of increasing amounts of energy was the measure of civilization. The anthropologist Leslie White posited four stages in human history: hunting and gathering, agriculture, industrial steam-power, and the atomic age.[18]

In the same years, Vannevar Bush urged Americans to see outer space as a new frontier that could be conquered through corporate research and development. In his vision, exploration and the search for new knowledge expressed the spirit of liberal capitalism. Profits, material progress, and science were inseparably joined. The march of science into the unknown would produce prosperity. In this "commodity scientism"[19] the space program was justified by improved commodities it "spun off" as by-products, such as new food concentrates, Teflon, and computer miniaturization. The *Los Angeles Herald-Examiner* editorialized: "America's moon program has benefited all mankind. It has brought better color television, water purification at less cost, new paints and plastics, improved weather forecasting, medicine, respirators, walkers for the handicapped, laser surgery, world-wide communications, new transportation systems, earthquake prediction systems and solar power."[20]

If during the 1960s the space program promised a technological cornucopia of goods, computer innovations played the same role in later decades. Full computerization appeared to promise a "new economy" that assured a permanent bull market on Wall Street. To some, the computer promised a virtual transcendence of not only economic law but of the natural world. In 1994, the Progress and Freedom Foundation declared: "The central event of the 20th century is the overthrow of matter. In technology, economics, and the politics of nations, wealth—in the form of physical resources—has been losing value and significance. The powers of mind are everywhere ascendant over the brute force of things."[21] The Progress and Freedom Foundation's cyberlibertarians asserted that computerization signaled a fundamental shift in the relationship between people and the natural world, because cyberspace fostered a new set of relationships in a virtual ecological system. Calling cyberspace "a bioelectronic environment that is literally universal," they compared it to the unknown world that explorers faced during the Renaissance. "The bioelectronic 'frontier,'" they suggested, "is an appropriate metaphor for what is happening in cyberspace, calling to mind as it does the spirit of invention and discovery that led ancient mariners to explore the world, generations of pioneers to tame the American continent and, more recently, to man's first exploration of outer space."[22] To them, cyberspace was a dramatic new chapter in the history of technological progress. (They seemed unaware that the American West c. 1800 was not empty space and that its settlement by Europeans displaced native peoples and transformed an indigenous ecological system.[23] Similarly, investment in cyberspace might be regarded not only as pioneering in a new space, but also as the displacement of resources from people to machines.)

Technological liberals note that more people are alive today than ever before. The total possessions of this population have been increasing, because for several centuries industrialization has multiplied humanity's command of productive power. These trends seem likely to continue. More engineers and scientists than ever before in history are at work on new patents and applications. Several billion people still live in poverty, but technological liberals expect that wider adoption of new technologies will lift them up. Though environmental mistakes have been made that needlessly destroyed much farmland and eliminated some species, they believe people can be more responsible without halting material progress. Most elected public officials share these liberal assumptions, which also underlie every corporation's statement of earnings.

Since the Renaissance, Western societies have been particularly adept at exploiting technologies to produce a surplus of food, goods, and services. They have used new forms of transportation to breach geographical barriers and integrate most of the world into a single market, collapsing space and time. In the long term, might this process lead to impoverishment? For example, building dams and irrigation systems in desert areas can increase food production, but after a few generations the land may become polluted and unproductive because of salts and chemicals left behind by evaporating water. Likewise, burning coal produces not only electricity but also smoke containing sulfur dioxide that falls to the earth as acid rain, destroying plants, fish, and wildlife. When such environmental effects are taken into account, the industrial revolution may only create temporary abundance. Another 100 years of intense use of fossil fuels will accelerate global warming, increase desertification, and cause many coastal areas to be flooded by rising seas.

From ancient times, some have regulated or even resisted technologies. A Byzantine city in the 530s had zoning laws that separated kilns, blacksmiths, and polluting activities from shops and houses.[24] Medieval French slaughterhouses and tanners polluted rivers and streams, leading to legislation in 1366 to prohibit such pollution of the Seine in Paris.[25] In the same years, demand for wood stripped much of England and France bare of forests. Firewood was burned for cooking and heating, and wood was also the primary building material. As early as 1140 the French had difficulty finding 35-foot beams for building, and architects worked out ingenious ways to use shorter pieces of wood to construct bridges and churches. Iron smelters consumed 25 cubic meters of wood to produce 50 kilograms of iron. A single furnace operating for just 40 days devoured an encircling forest for a one-kilometer radius. By 1230 the English were importing wood from Norway, and English coal was sold not only in London but also on the Continent. Mines became ever more extensive, and by the end of the thirteenth century air pollution was a problem. In 1307 London was the first city to prohibit coal burning, in a royal proclamation that was generally ignored. Many contemporary ecological problems—deforestation, fuel shortages, and both air and water pollution—can be traced back in European history at least 700 years.

As Europeans ran short of raw materials and expanded into the rest of the world, these problems recurred wherever they went. Their demand for wood, charcoal, and iron stripped colonies of forests. European demand for gold, silver, copper, zinc, lead, and nickel created extensive mines and open pits, immense slag heaps, and polluted groundwater. European industrial methods, whether exported to colonies (South Africa, India) or voluntarily adopted by other countries (Japan), led to extensive air and water pollution. Europeans' farming methods, exported to their colonies, brought new areas into production but also dramatically accelerated soil erosion. Plowing land and letting it lie fallow were not destructive practices when applied to the heavy soils of Northern Europe, but they often proved catastrophic in drier regions with lighter soils, such as the western plains of Canada and the United States. In Latin America, Europeans introduced new grazing animals whose hooves loosened the soil in hilly and mountainous areas, accelerating erosion.[26] More recently, intensive use of pesticides and fertilizers has increased agricultural yields, but often at the cost of poisoning the soil and the groundwater. In short, Western technologies have been used to create abundance, but at a high environmental cost. In the twentieth century alone, the United States lost to erosion topsoil that took 1,000 years to form, and it continues to lose topsoil at a rate of 1.7 billion tons a year. In the world as a whole, agricultural land seven times the size of Texas has been destroyed through erosion and misuse. The UN estimates that between 0.3 and 0.5 percent of the world's cropland is destroyed each year, creating pressure to clear more forests, causing yet more erosion.[27] Clearly, these are grounds for pessimism.

Technological pessimism was prominent in Great Britain, where the industrial revolution began. William Blake denounced the "dark Satanic mills" of the English Midlands, and William Wordsworth complained about a railroad built into his beloved Lake District ("Is there no nook of English ground secure / From rash assault?"). And in *Frankenstein* (1818), a novel still resonant today, Mary Shelley evoked the possibility that scientists might create monsters that would escape their control. In contrast to today, during the nineteenth century, the attack often came from the political right. As Maxine Berg notes, the magazines *Fraser's* and *Blackwood's* published "the Tory theoretical attack on industrialisation and its social effects." Resistance to mechanization did not merely express a Tory "bias against the middle class" but was "a deep protest against the whole mechanism of industrial society."[28] Thomas Carlyle called the nineteenth century "the age of machinery." He declared that men had grown mechanical, and that steam had "fearfully accelerated a process which was going on already, but too fast."[29] Society had become "a huge, dead, immeasurable steam engine, rolling on, in its dead indifference."[30] "Were we required to characterize this age of ours by any single epithet," Carlyle mused in 1829, "we should be tempted to call it, not an Heroical, Devotional, Philosophical, or Moral Age, but above all others, the Mechanical Age." Carlyle's critique also included the machine's psychological effects: "Men are grown mechanical in head and in heart, as well as in hand. They have lost faith in individual endeavour, and in natural force of any kind. Not for internal perfection, but for external combinations and arrangements, for institutions, constitutions—for Mechanism of one sort or other, do they hope and struggle."[31]

Critics on the right and on the left took up Carlyle's rhetoric. One can trace a tradition from such nineteenth-century pessimism to Henry Adams, who judged the automobile to be "a nightmare at one hundred kilometres an hour, almost as destructive as the electric tram which was only ten years older."[32] Adams concluded that technology as a whole had accelerated out of control.[33] After the appalling destruction of World War I, his autobiography sold briskly and helped convince intellectuals and writers of the 1920s that technological civilization had produced what T. S. Eliot called "the waste land." Eliot's generation had never heard of global warming, but by the year 2000 "waste land" was not a metaphor but a description. With half a billion automobiles on the world's highways, air pollution drove up the level of greenhouse gases. Furthermore, making cars requires enormous resources. Producing, gathering together, and assembling all the parts that go into a typical car—steel, plastic, aluminum, rubber, glass, and so forth—requires as much energy as it does to drive that car for a decade. Furthermore, roads, driveways, parking lots, and service stations have now expanded to cover more than 5 percent of the land in industrial countries. Nor are cars a particularly safe form of transportation. At the end of the twentieth century, traffic accidents annually killed 400,000 people and maimed many more.[34]

There were other, less visible problems. In *Silent Spring* (1962), Rachel Carson warned that the abuse of pesticides such as DDT had poisoned many areas and undermined their ecological systems. An ever-larger chorus of critics identified advanced technology not only with increasing numbers of automobiles and air pollution but also with atomic bombs, chemical pollution, biological mutants, malfunctioning computers, and out-of-control technical systems. These dystopian fears recurred in science fiction, which often depicted a devastated future, where the remnants of humanity struggled to survive the wreckage of a technological disaster. At the beginning of the twenty-first century, people could use new technologies to increase production, but a growing minority doubted the wisdom of the goal.

Even if, despite pollution and overuse of resources, people have increased abundance for all, are they happier as a result? According to some polls, even though the gross national product had doubled, Americans of the 1990s were no happier than they had been in 1957.[35] Their work hours had increased, while time spent with friends and family had declined. Bombarded with thousands of advertisements a day, they overconsumed in a throwaway economy.

Should desire for more and more things drive human development? In conceiving the first modern utopia, Thomas More rejected high consumption. His Utopia increased leisure by drastically reducing human wants and adopting a modest style of life.[36] Utopia's citizens rejected luxury on principle. They wore simple, long-lasting clothing, and lived in small, sturdy houses. More had "little confidence in tools or practical arts either as emancipators or as promoters of social equality."[37] Likewise, Defoe's Robinson Crusoe soon realized that there was no point to killing more animals or growing more food than he could consume: "I had enough to eat and to supply my wants, and what was all the rest to me?"[38] Indeed, when Crusoe salvages bars of gold from a shipwreck, they are of far less value to him than a fire shovel and tongs.[39]

In contrast to More and Defoe, however, many utopian writers of the nineteenth and twentieth centuries projected a future world with ever-increasing levels of luxury. In Edward Bellamy's *Looking Backward,* which sold more than a million copies in the 1880s and which remains in print today, a centralized state controls a disciplined "Industrial Army" that mass-produces a cornucopia of goods. Each citizen has the right to an ample supply from the large warehouses that have replaced wasteful, individual stores.

Western societies embraced the ideal of technological abundance, but a small, articulate minority called for simplicity. Henry David Thoreau argued that, rather than constantly expand one's desires, it was better to simplify material life to make time for reading, reflection, and close study of nature. In *Walden* (1854) he ridiculed the farmer who spent his life acquiring more possessions, arguing that such a man had lost control over his life. More generally, Thoreau questioned the value of slicing life into segments governed by clock time and suggested that the railway rode mankind rather than the reverse. He concluded that "men have become the tools of their tools," and feared that "our inventions are wont to be pretty toys, which distract our attention from serious things. They are but improved means to an unimproved end.[40] . . . We are in great haste to construct a magnetic telegraph from Maine to Texas; but Maine and Texas, it may be, have nothing important to communicate."[41] Though these remarks on the telegraph may seem quaint today, in point of fact Americans did not immediately know what they might use the telegraph for, and it took the better part of a decade before the first intercity line was completed (from Washington to Baltimore, in 1844). At first this line had few customers; in 1845 it operated at only 15 percent of its capacity.[42] To stimulate public interest, the promoters staged long-distance chess games.

Even people who have never read Thoreau's *Walden* know that it is based on his experience of building and then living in a simple one-room house in the Massachusetts woods. Thoreau sought to reduce his needs to a minimum, but he was by no means anti-technology. Beginning in the 1820s, Thoreau's family manufactured pencils, a process that involved the careful adjustment and use of machinery. Thoreau invented a machine to grind plumbago (graphite) into a uniformly fine powder. He then discovered how to combine this powder with a particular clay to make a "pencil lead" that wrote with an even, smooth line.[43] The Thoreau family also sold finely ground plumbago to printing companies. Thoreau was skilled enough in the use of the axe and the hammer to construct his modest house. He was interested in how sawmills and gristmills operated, and his journals contain descriptions of work of many kinds, including the harvesting of ice from Walden Pond. He made a small cash income practicing as a surveyor, measuring the land scientifically by means of repeatable processes that were subject to verification. Far from being a technophobe, Thoreau had many of the traits of the Yankee mechanic. In Concord, he got along better with skilled workers than with the gentry.[44] He described himself to his former Harvard classmates as follows in 1847: "I am a Schoolmaster—a Private Tutor, a Surveyor—a Gardener, a Farmer—a Painter, I mean a House

Painter, a Carpenter, a Mason, a Day-Laborer, a Pencil-Maker, a Glass-paper Maker, a Writer, and sometimes a Poetaster."[45]

Once Thoreau had completed his little house, he selected furnishings, including a bed, a table, and so forth. Initially, he had a little rug for a doormat. He soon found that this small rug had to be taken up and shaken, and that the floor beneath still needed cleaning in any case. The rug was merely a small nuisance, and he got rid of it, because he wanted to eliminate, rather than accumulate, such possessions and the little tasks they entailed. He realized that people can easily become slaves to what they own. Likewise, he concluded that it was often faster to walk than to ride a stage or a train, if one took into account not just travel time but also the number of hours one had to work to earn money for the fare. By Thoreauvian logic, a good many conveniences not only prove unnecessary; they create debt and force us to work long hours so that we can pay for them.

Thoreau's life and writings inspired other early environmentalists. John Muir advocated simple living arrangements and questioned the value of a clutter of technological gadgets. Others who have retreated from civilization into remote places to practice simple living include Helen and Scott Nearing, who left New York to take up subsistence farming in the 1920s. For several generations they proved that one might have a good life with a minimum of technology. During the twentieth century, thousands of others also quietly rejected technological abundance, as documented in David Shi's book *The Simple Life*.[46] Not all of them retreated into the countryside. Lewis Mumford advocated a simple life in the city, sought to establish planned communities, and attacked the consumer culture as "the opulence of carefully packaged emptiness." In the last three decades of the twentieth century, inspired by such individuals and by books such as *The Poverty of Affluence*,[47] non-profit political organizations such as World Watch and Friends of the Earth lobbied against the ideology of growth. They raised questions about scale and appropriateness that seemed urgent once the energy crises of the 1970s revealed the vulnerability of Western economies to oil and gas shortages. In *Small Is Beautiful* (1972), E. F. Schumacher advocated that developing countries solve their problems with simple machines and small-scale workshops rather than complex, high-tech "solutions" that made them dependent upon foreign aid and imperial suppliers.[48] Such "low-tech" ideas also seemed appealing in industrial nations, because they empowered individuals to select and construct their own technological systems. The counterculture found such ideas especially appealing and codified them in *The Whole Earth Catalogue* (1968), which provided detailed information on where to buy and how to use a host of small-scale technologies to preserve food, generate power, build and insulate new forms of shelter such as geodesic domes, and otherwise escape from conventional consumption.

With the energy crisis of the early 1970s, people other than counterculturists became interested in such things as passive solar heat and electricity-generating windmills. Suddenly it seemed self-evident that oil, gas, and coal were limited resources that would run out within a few generations. Nor was the use of nuclear reactors to generated electricity considered a comfortable alternative. Even before the accidents at Three Mile Island and Chernobyl demonstrated the dangers of nuclear plants, Amory Lovins attacked nuclear power generation as a brittle, centralized system that was costly, created long-term pollution, and was vulnerable to terrorism. In contrast, wind and solar power were flexible, de-centralized, non-polluting, safe, and probably less expensive in the long term.[49] In the short term, however, alternative energy sources could not begin to supply the electricity demands of Western economies. Furthermore, so much energy was needed to produce some of the early solar panels and windmills that they represented little net gain.

Beginning in the 1980s, World Watch issued yearly "State of the World" reports. Conventional economic theory assumes that the gross national product must continually grow, but such groups were convinced that technological abundance both destroyed the environment and distracted people from helping one another.[50] Such reports and the ideal of sustainability based on alternate energies did not transform the habits of most people, however. Once the energy shortages of the 1970s receded, consumers reverted to automobiles for transportation and continued to increase domestic per-capita energy use. After 1980, automakers found that Americans did not want small, fuel-efficient cars; they wanted pickup trucks and so-called sport-utility vehicles. Just as Henry Ford discovered in the 1920s that the public would not buy the Model T in perpetuity but demanded annual restyling, in the last two decades of the twentieth century manufacturers acceded to consumers' demands for bigger vehicles.

In contrast, Europeans have long been accustomed to smaller automobiles and higher taxes on gasoline, which encourage alternative forms of transport. Many rely primarily on mass transit and bicycles and live in compact cities with many row houses and apartments. Such countries have a standard of living as high as in the United States, but use only half as much energy per capita. Their economies may prove more sustainable than those of Australia, the United States, and Canada. The Netherlands is a particularly interesting example of a highly technological nation that decided to limit development. For centuries, the Dutch approached the natural world as a reservoir of raw materials to be exploited to the maximum. They drained low-lying areas and made them into fields. They transformed rivers into canals. They built dikes that pushed back the ocean, and today much of the population lives below sea level. However, in the twentieth century the Dutch began to recognize limits to this exploitation. After cutting off the Zuider Zee from the North Sea, they decided not to drain all the water away. They recovered some land for settlement and agriculture, but allowed much of the area to remain under water. They had prevented North Sea storms from flooding the heart of their nation, but long experience had shown that pumping out an entire area provoked land subsidence, so that the new fields sank below the original seabed. Simultaneously, the Dutch learned to see the tamed Zuider Zee as the equivalent of a giant lung in the middle of the country, providing moisture, cleaner air, fishing, ecological diversity, and recreation areas.

But during the centuries when European cultures conquered much of the rest of the world, often with destructive results, they imposed few checks on development. North American passenger pigeons, once so numerous that they seemed to blot out the sky, were extinct by the late nineteenth century. These migrating birds had once eaten billions of insects each year and fertilized the landscape with tons of droppings containing phosphorous. Their extinction broke an important part of an ecological cycle.[51] Restrictions on hunting the carrier pigeon and preservation of its habitat could have saved it, but no one realized the bird's importance until too late. More recently, some European nations have reversed a destructive policy of eliminating hedgerows to create larger fields as part of the "rationalization" of farming. Many bird species live and prosper in hedgerows, and as they diminished local ecologies were upset. By the end of the twentieth century, hedgerows were being re-established to protect against wind erosion and to readjust the balance between wild nature and agriculture.

The refusal to drain the Zuider Zee and the reestablishment of hedgerows suggest that humanity can move beyond seeing technology and nature as irreconcilable opposites. Richard White has written of the Columbia River Valley as an "organic machine" whose energies include not only the flow of the water, but also the salmon, the Native Americans, and the Anglo-Americans, with their steamboats, dams, and electrical systems. White rejects the man/nature dichotomy, and sees an intermingling. "Dams, hatcheries, channels, pumps, cities, and ranches are all products of human work, and it is our labor that ultimately links us to the river. Our labor, our energy, is the nature in us. And we harness it, just as we harness nature, to social purposes."[52] Many environmental scholars now see people as part of nature, and nature as part of culture. If instead one views nature as something separate from humans, it creates a misleading perception of "an 'us vs. them' posture that creates opposition within environmental movements as well as outside them."[53] An office worker may feel quite independent of the natural world, forgetting that the electricity driving the computer and lighting the building comes from a dam in the mountains or from coal mined and burned hundreds of miles away.[54] In larger perspective, human and natural processes are inseparable.

Joel Cohen confronted the fundamental question in his 1995 book *How Many People Can the Earth Support?* For more than 300 years, analysts have argued there must be a limit to the earth's long-term "carrying capacity," but they have disagreed about what it is. Estimates have ranged from 1 billion persons to over 100 billion. The medians for low and high estimates suggest a maximum population of between 7.7 billion and 12 billion.[55] Earth's population is now moving into this range, and theories about its carrying capacity will be tested. Aside from different statistical assumptions and varying estimates of the available arable land and potable water, "how many people the Earth can support depends on what people want from life."[56] If people in dry areas want green lawns and chlorinated swimming pools, there will be less water for irrigation. Not only will farmers produce less food, but chlorinated water will be unavailable to other species. If people want to eat meat every day and wear natural fibers, the world can support fewer people than it can support if people are vegetarians and buy synthetic clothing. Ultimately, the world's carrying capacity is not a scientific fact but a social construction. Nature is not outside us, and it does not have fixed limits. Rather, its limits are our own.

Does technology assure abundance? The example of Robinson Crusoe creating a comfortable life on a remote island is more complex than it first appears. When he is leaving his island, he contrives to settle several sailors there, and he sends them cattle, hogs, and additional settlers. Crusoe transforms himself from a castaway into the owner of a colony. Using European agriculture, metal tools, and weapons, the settlers increase the island's carrying capacity and link it to the Atlantic economy. Its limits depend on how much its inhabitants want. That island's story, like humanity's, is open-ended, depending on human choices.

Work: More, or Less? Better, or Worse?

Much hard physical labor has disappeared, as workers use machines to dig canals and ditches, pump water, carry materials from place to place, and lift loads in factories and warehouses. Today, in the industrial nations, few field workers swelter in the sun; agriculture has been mechanized. But some jobs are numbingly repetitive: the supermarket cashier endlessly scanning foods at the checkout counter, the fast-food cook frying identical hamburgers, the toll-booth attendant collecting small sums from hundreds of passing motorists. Such work is boring, it isolates workers from sustained contact with other people, and it does not lead to new opportunities. Some of these jobs are being automated. Many motorists no longer need to stop to pay tolls, as scanners read the codes on their vehicles. Tedious and dangerous factory work has been replaced by robots or automatic machines, while millions of people hold jobs that did not exist 150 years ago in industries created around new technologies, such as computing, musical recording, broadcasting, design, advertising, and research and development. As recently as 150 years ago, most people in Europe and the United States were farmers. Today, less than 5 percent remain on farms, and industrial work occupies only about 25 percent of workers. In agriculture and industry, technological efficiency has meant more production from fewer people. This efficiency at first may seem to be unquestionably a good thing.

Only three centuries ago, people made almost everything by hand, using relatively simple technologies. Weavers produced cloth on their own looms. Shoemakers, tinsmiths, coopers, bakers, candlemakers, carpenters, glass blowers, printers, and hundreds of other artisans each knew all the intricacies of a trade, and made goods from start to finish. Their knowledge was for the most part not written down but literally handed down from master to apprentice. The tools changed little from one generation to the next. Indeed, a museum near Barcelona displays a wide range of tools from the Roman Empire, many of them still familiar. Much that workers knew was literally in the hands and the body, and a master worked with speed and grace that novices struggled to imitate. A technology consists of both tools and skills.[1] Cooking is an excellent example. Recipes provide outlines sufficient only for the experienced, and, as the popularity of cooking programs on TV attests, it helps to watch someone else.

Work is a social practice, requiring coordination between people, and a workplace contains an ensemble of tasks that must be orchestrated. Managers or foremen impose some of the coordination, but some of it must arise from the workers themselves. Arnold Pacey has written perceptively about the subtle rhythms of physical work, noting that a scythe, a hammer, or some other tool is used most effectively in an essentially musical way that has to do with breathing, pacing, and patterns that can best be learned from other practitioners.[2] Hunters also had musical traditions to make it easier to hunt animals.[3] Field workers and sailors often sang while working, capturing in songs the rhythms required for harvesting, pulling, rowing, hoisting, and many other tasks. In some jobs, such as mining, getting the rhythm right is a matter of life and death, as acting against the rhythm puts a worker in harm's way.

The factory system undermined the musical aspects of traditional work. Steam engines had rhythms, too, but they were not tuned to the movement of the body. Factories also did away with the artisan's comprehensive knowledge of how to make something from start to finish, substituting specialized machinery and the subdivision of labor. Weaving was far less a matter of skill after thousands of handloom weavers lost their livelihood and their crofts and the tasks of replacing bobbins and mending broken threads on automatic looms were given to women and children. Labor historians often speak of the "de-skilling" of workers, and this may seem to be the main result of the creation of factories. Yet a small number of ingenious mechanics and machine builders received higher wages, because they became essential to constructing and improving the new factory production systems. The coming of industrialization meant several different things: unemployment for some skilled artisans, monotonous low-wage work for others (often women or children), high wages for a few mechanics, and some new jobs in the factory hierarchy (for example, in marketing and accounting). The adoption of the computer in workplaces seems to be having similar effects.

Industrialization did not create a permanent underclass; rather, it shifted factory laborers to white-collar work. But this shift took place over many decades, and for generations of workers the factory system meant alienation. Their artisanal knowledge was undermined, they no longer owned the tools of their trade, and they lost their identification with the workplace and its products.

As factories became more productive, manufacturers could choose how to spend the surplus. They could raise wages for workers, lower prices for consumers, take greater profits for themselves, make further improvements in the machinery, or allocate resources to all of these options. They often focused on low prices, not only to win market share but also to drive remaining artisans out of work, completing the process of industrialization. Most managers and owners also invested some of the surplus in new machinery, and they all took profits. They were less likely to pay higher wages, though there were a few exceptions (such as the utopian industrialist Robert Owen).[4] Usually, workers had to organize into unions and to strike or take industrial action in order to get their share. Confrontations and strikes gradually established higher wages, safer workplaces, and some pension benefits.

In telling this story, economic and labor historians often focus on wages, benefits, prices, and profits, while treating the system of production as a "black box." They take the machine's greater productivity as a given. In contrast, historians of technology open up that box to see how the machines work, to ask what technical choices were possible, to ask what skills were replaced and what new skills were demanded, and to find out how particular workers organized particular machines into a production system. Merritt Roe Smith's study of gun manufacturing in the years preceding the American Civil War exemplifies this approach.[5] By studying the actual work practices at the federal armories and examining the guns produced, Smith showed that mass production was neither as easily achieved nor as early accomplished as most people had assumed. Historians had conventionally given Eli Whitney credit for manufacturing using interchangeable parts as early as 1801, but for decades the parts coming from the molds had to be filed and fitted. Skilled workmen remained at the core of gun production throughout the antebellum period, and their organization and productivity varied considerably in different arsenals.

Smith found that Whitney never quite achieved his goal of inter-changeability, which was only reached half a century later. Many early factory workers felt that the tradeoffs of industrial-ization were not worth it. Factory labor robbed them of customary freedoms. Where intense bouts of labor once alternated with spontaneous breaks, gradually the arsenals, textile mills, and other factories imposed repetitive work at an unvaried pace. Where once workers had decided the order, the flow, and often the cost of the work, as well as exercising informal quality control, factory managers wanted to dictate these matters with little consultation. This seems to have been particularly true of managers with engi-neering degrees. For example, when engineers installed air condi-tioning in factories they used it to increase their authority and control. They first insisted on control of the configuration of the building and then over the activities within. In the food and brew-ing industries the new technologies of automatic temperature control and air conditioning often replaced or de-skilled workers, whose tacit knowledge of materials and sensitivity to different atmospheric conditions had been crucial to producing such var-ied things as macaroni, candy, beer, and cigarettes.[6] In hundreds of ways, managers sought to reorganize, rationalize, and control the factory. Workers lost control of their time, their space, and their movements, and grudgingly gave up some of their tacit knowledge of materials and processes. Music also went out of the work. Today's worker is far less likely to harmonize with others than to listen to recorded music on a headset, blocking out the sounds of the workplace. It is little wonder that generations of workers rebelled against the factory, rejecting not only the long hours and low wages but also the monotonous routine and the loss of autonomy. After c. 1900, strikes seem to be primarily about wages, but in the eighteenth and nineteenth centuries workers often struck over the rigid time discipline that machines imposed, restrictions of movement about the factory floor, or the loss of decision making to foremen or engineers.[7]

American apologists for slavery lambasted the exploitation of the working class and idealized the paternal responsibility of slave owners. Slavery's defenders attacked cotton mills, arguing that free labor was more profitable to the capitalist than slave labor was to the planter because the capitalist took no social responsibility for his workers. Drawing on August Comte, Henry Hughes's *Treatise on Sociology* found the slave system superior to industrialism because it guaranteed workers employment, housing, medical care, and religious instruction. When hard times came in the industrial North, Hughes argued, the workers starved. When a mill worker was sick, old, or injured, the capitalist simply hired someone else. He had no interest in keeping her healthy, and lost nothing if she died young.[8] Hughes's apologia for slavery may seem an unreliable critique of the factory system, yet much the same criticism of industrial society recurred elsewhere. Alexis de Tocqueville, after studying the new society of the United States in the 1830s, contrasted the sense of social obligation felt by a feudal aristocracy with the lack of any obligation beyond explicit contracts in modern industrial society: "The territorial aristocracy of former ages was either bound by law, or thought itself bound by usage, to come to the relief of its serving-men and to relieve their distresses. But the manufacturing aristocracy of our age first impoverishes and debases the men who serve it and then abandons them to be supported by the charity of the public."[9]

Workers faced increasing pressures toward conformity and control even in good times. Frederick Winslow Taylor's *Principles of Scientific Management* (1911) epitomized the trend.[10] Taylor argued that there was "one best way" to do any job. He broke down any given task into discrete actions, and then taught workers the most efficient sequences and movements and the correct pace for even the simplest activity, such as shoveling. He also invented or modified tools, such as shovels of different shapes and sizes to handle materials of different weight. He organized individual tasks into a rational sequence, so that work flowed evenly. Those who cooperated produced more. In return, Taylor expected managers to raise workers' pay, though not all employers remembered this part of his system. His book sold widely in French, German, and Russian translations, and Lenin praised it. But American labor leaders loathed Taylor's approach because it took agency away from workers. When implemented, his system often provoked strikes. Nevertheless, "Taylorism" had an enormous resonance beyond the factory in many areas, including home economics, education, and popular culture. Experts appeared in every area of life, proclaiming that they had discovered the "one best way" to lay out a kitchen, regulate traffic flow, or plan a community. In the neo-Marxist thought of Antonio Gramsci, Taylorism was an oppressive historical stage in a sequence that had begun with primitive accumulation and would end with the collapse of capitalism.[11] In practice, however, Taylor was but one of many "experts" who attempted to dictate the rules for more efficient working and living.

Historians long overestimated the degree to which workers were de-skilled and reduced to mere cogs in the factory machine. Recent studies have emphasized the survival of elements of craft labor beyond the 1890s, as well as the persistence of worker control on the shop floor. Because many industrialists realized that the "manager's brains are under the workman's cap,"[12] some workers continued to organize themselves as teams that contracted their services to managers. Such workers at times also decided how to divide the work and how much each should be paid.[13]

Thorstein Veblen, having observed the rise of scientific management, the expansion of the engineering profession, and the continuing desire among workers to find meaning and satisfaction through their jobs, thought an end to capitalist control of the machine was imminent. Only the new class of technicians seemed to understand modern productive systems well enough to use them efficiently. From Veblen's perspective, the capitalists were not competent to control the system, and the best form of government therefore would be a "technocracy," in which factories would be run by those who understood them. Such ideas found formal expression in the technocracy movement, and later inspired some New Deal projects of the 1930s.[14]

Even as scientific management spread, Henry Ford's engineers developed the assembly line, which after 1913 quickly spread to many other companies. This form of continuous-flow production drew on experience in the arms industry, slaughterhouses, and many other areas, but it was considered to be such a startling innovation that thousands of tourists visited Ford's factories to see it in operation.[15] Taylor had retrained the worker to do the same job more efficiently; the assembly line completely redefined the job. Taylor offered workers incentives to do more piecework; Ford made piece rates pointless, because the assembly line paced the work. Instead, workers received a high fixed wage. Taylor had designed ideal tools, such as different shovels for materials of different weight; Ford abolished shovels in favor of electric cranes, moving belts, and other aids to continuous-flow manufacturing. But although the assembly line radically reduced the time needed to assemble a car, workers found the repetitive labor mind-numbing. As factories adopted mass production, labor turnover rose to 10 percent per month—that is, more than 100 percent per year.[16] Workers were voting with their feet against hierarchical rigidity, against the accelerating pace of work, and against the management's control of the shop floor. (Fortunately, for some a rapidly growing white-color sector offered an alternative.)

Mass production characterized many large firms from 1913 until the 1970s, when Japanese automobile companies pioneered lean production. That style of production no longer gave single tasks to single workers; it gave teams of workers sets of tasks and more say in decision making. The movement of the assembly line no longer entirely dictated the pace, as workers were expected to stop the line if necessary in order to complete their tasks. Thus, mistakes or quality problems were fixed immediately rather than being passed along. This change went hand in hand with an innovation in inventory handling. Instead of keeping large quantities of spare parts on the factory floor, backed up by nearby warehouses, improved communications and transportation were used to ensure "just-in-time" delivery. In effect, this made suppliers produce and deliver at precisely the pace demanded by production. Corporations thereby saved a fixed expense of millions of dollars that were formerly tied up in spare parts. By 1987, in half the factory space and with half as many workers, Toyota could assemble cars twice as fast as General Motors. These cars were designed with fewer total parts and contained only one-third as many defects.[17]

American factories had to adopt the new methods to stay competitive on both quality and price. Management loved lean production in theory because it decreased investment in plant space and wages while increasing profits. One General Motors plant in New Jersey adopted elements of lean production in the late 1980s, and the changeover eliminated one-third of the production workers and 42 percent of foremen and supervisors.[18] Many workers lost their jobs through buyouts. Those who remained had an ambivalent response. The work was less monotonous, but the greater responsibilities and the increased tempo were stressful, especially because American managers often objected when individual workers stopped the assembly line (as they were supposed to do) in order to fix a problem. The line worker trying to ensure quality control ran afoul of management's desire for maximum throughput.[19] Furthermore, though the introduction of robotics and computer-driven machines more than doubled the demand for highly skilled workers, these still constituted only 12 percent of the total workforce. The remaining workers may have been organized into teams and given more responsibility, but the work remained repetitive. Not surprisingly, the highly skilled workers were the ones most satisfied with the new production system.

For at least 200 years, new technologies such as the assembly line and new organizational systems such as Taylorism, just-in-time parts delivery, and lean production have eliminated jobs. Who has not heard of someone made redundant by a new machine or a new process of production? In the era of "silent" films, professional musicians had regular employment in the movie theaters. No more. Most telephone operators have disappeared, replaced by automatic switchboards. Many bank tellers have disappeared, replaced by automated teller machines. Computer programs have replaced many white-collar workers, allowing Internet users to file forms, pay bills, and make applications online. Will all the jobs disappear? In the 1990s Jeremy Rifkin sounded the alarm: "The transition to a near-workerless, information society is the third and final stage of a great shift in economic paradigms. . . ."[20] When agriculture mechanized, the surplus farm workers went to factories. When factories introduced assembly lines and labor-saving machines, blue-collar workers and their children moved into office work and services. "In the past, when new technologies have replaced workers in a given sector, new sectors have always emerged to absorb the displaced laborers. Today, all three of the traditional sectors of the economy—agriculture, manufacturing, and service—are experiencing technological displacement, forcing millions onto the unemployment rolls."[21] The prospects indeed seemed dim, yet when Rifkin's book appeared in the 1990s, the American economy was creating an average of a million new jobs per year. The rapid computerization of the economy was not producing the dire effects Rifkin predicted.

Rifkin was by no means the first to misread the signs. From the eighteenth century on, many workers feared technological unemployment, from the handloom weavers of early industrial Britain, who were made redundant by mechanical looms, to the white-collar office workers of the 1990s, made redundant by computer software. In England, such concerns once prompted the machine-breaking raids of the Luddites. Workers in every industrializing country have protested the transfer of their skilled work to machines. In recent decades, typesetters, steelworkers, and meat packers have lost their jobs to new machines and new production techniques. For generations, witnesses to such change have warned that consumption could not keep up with production, and that the result would be high unemployment. An 1878 American pamphlet warned that the saturation point had been reached and called for shorter working hours for all so that the scarce work could be distributed evenly. During the 1930s, technology seemed the central cause of unemployment and overproduction, and one critic declared that, because there were limits to what people could consume, industrial society had reached a condition of permanent stagnation. Because mass production had both lowered the cost of goods and increased their quality, consumer demand would fall and the number of jobs would decline.[22] In 1933, a similar logic led the US Senate to pass a bill calling for a 30-hour workweek, with the goal of redistributing work to the 15 million unemployed. Instead, President Franklin Roosevelt created make-work programs.

In more ominous visions, machines replace workers altogether. In Fritz Lang's 1927 film *Metropolis,* society is divided sharply between the rich and the poor. The latter live and labor underground, beneath the shining towers of the city. A master technician finds a way to replace them with robots. The wealthy then provoke a rebellion so that the workers can be crushed and eliminated.

From about 1930 to 1980, amid such fears and predictions, Americans conducted an intense debate over technological unemployment.[23] Manufacturers and engineers consistently argued that technology is the source of progress, asserting that job losses are only temporary displacements. Most jobs eliminated are repetitious and menial, while some of those created, such as machine building or computer programming, require imagination and skill. Many workers are only out of jobs temporarily. The fault for the Great Depression, in this view, lay with incompetent economists and politicians, who had not yet learned how to make the most of advancing technology. Engineers were particularly attracted to the "cultural lag" theory, perhaps because it shifted the blame for unemployment away from them to the larger society. When the 1939 New York World's Fair depicted an idealized world of 1960, it avoided any suggestion that machines would displace workers. Instead, it assured the public, as a Westinghouse film asserted, that corporate research was creating "so many jobs, there won't be enough people to fill 'em."[24]

Nevertheless, the fear remained that assembly lines, robots, and new machines would create massive unemployment. Kurt Vonnegut, in his 1952 novel *Player Piano,* describes a future in which, as in Lang's *Metropolis,* machines do all the work, supervised by a few technicians. But Vonnegut did not imagine the liquidation of the working class. Rather, he depicted a future in which most people are idle, purposeless, and powerless.[25] Such fears were especially common after World War II. In the postwar years, the US government invested in computer-aided design (CAD) and computer-aided manufacturing (CAM). CAD-CAM eventually made it possible for the designer's representation of a product on a screen to be manufactured directly from computer-generated instructions. It sought to incorporate workers' skills and routines into standard machine operations. Vonnegut had already described the goal: all worker knowledge would be translated into mathematical formulae that the computer could use to replicate what they once had done. Technicians would replace people with robots doing routine tasks, and computers would monitor the behavior of the few workers remaining.

A tooling engineer explained that the goal of numerical control was "to go directly from the plans of the part in the mind of the designer to the numerical instructions for the machine tool." By eliminating workers this increased productivity. "Since the rate of production of such a machine [tool] is several times that of a conventional machine, work is effectively taken away from at least two or three skilled machinists."[26] Yet automation and CAD-CAM also created new niches. Highly skilled workers were needed to set up and maintain the new "automated" machines, and savings were less than anticipated. In practice, it is "impossible to dispense with a core of skilled workers. Whether they are engaged in repairing the existing equipment or in installing the next generation of technology, they must be capable of understanding each task as part of a larger complex of tasks. . . ."[27] Such workers become more indispensable for "lean production" than some managers.

As global communications improved, the distance between designer and machine operator could increase. An engineer in California could prepare instructions for machine tools operating in Mexico or China. These developments seemed ominous to workers in Europe and the United States, as manufacturing shifted to offshore locations. Steel production, for example, increasingly took place in countries with lower wages and less exacting environmental standards. This "de-industrialization" worried many between 1979 and 1984, when in the United States alone factory shutdowns, relocations, and plant redesign cost 11.5 million workers their jobs.[28] The process continued during the next two decades, assisted by free-trade agreements with many nations. European workers saw industrial jobs disappear to their former colonies in Latin America and Asia. Not only was capital being substituted for labor; the remaining jobs were being exported.[29]

Yet the story had another side. As computers became more widespread, better software made them more user-friendly. It soon appeared that computerization could lead to unexpected results, because the textualization of factory knowledge applied to managers as well as to workers. During the 1980s, Shoshana Zuboff studied fully computerized workplaces in operation and saw that, despite management's explicit intention to use computers to de-skill workers and to increase its control of operatives, something quite different was taking place in some companies.[30] Computers demystified management, and skilled workers could use them to undermine hierarchy, secrecy, and centralization. A fully computerized plant became more transparent. Communication no longer flowed only from hidden sources at the top down to middle managers, who then permitted only a trickle of information to reach their subordinates. Instead, computers potentially gave every operator access to all the information in the system, and clever workers were quick to appreciate this fact, particularly if it could aid them in disputes. Previously, in order to protect themselves, both workers and managers withheld a good deal of their working knowledge from one another, preserving realms of autonomy. In part, they did this intentionally in order to preserve power, but in part procedures and assumptions were often difficult to verbalize. In contrast, at a fully computerized plant much of this unspoken working knowledge was transformed into text that provided more and better information about the work process.

As corporations adopted computers, they realized that middle management was less important than skilled workers. "Lean production" already had shown that when workers had more information and more scope they could find creative ways to improve both the product and production. Such ideas spread rapidly during the 1990s. One General Electric executive declared that "all the good ideas—all of them—come from the hourly workers."[31] The logical yet startling conclusion was to fire many middle managers and to forge cooperative ties with unions.

In this changing work environment, some unions conceive their proper role not as adversarial negotiators but rather as "value-adding organizations."[32] Rather than demand fixed job definitions and pay scales, these "value added" unions agree to flattened work hierarchies in exchange for giving workers increased decision-making powers. In some cases, union leaders sit on the corporate board and vote on such matters as where new factories will be built. At Harley-Davidson the partnership between the unions and management has gone so far that the company's president shares office space with the presidents of the two local unions. There are said to be "no walls, no partitions, no secrets,"[33] and the unions focus on management skills more than on formal bargaining.

In theory, mid-level managers who observe these trends might resist the empowerment of workers in a computerized factory, for example by inhibiting information flow in order to protect their jobs. But managers soon find that they, too, are only part of a larger hierarchy of observation and information. Each level must report to the level above. At sites ranging from sawmills to telephone companies, they soon discovered that it was difficult to restrict the flow of information between different levels of the organization. At this stage, "the adversarial vocabulary of 'us' and 'them' invaded the language of operators toward their managers, and plant managers toward divisional executives. This mistrust was not rooted in a perception of evil or malicious intent. It was, instead, the feeling evoked in the silent dance of the observer and the observed."[34] For managers and workers alike, when a computer system recorded every action and made it available for analysis, the result was "a sense of vulnerability and powerlessness." However, it was not prudent for one factory to turn back the clock and return to pre-computer operations, unless its competitors did so as well. Managers began to realize that they could develop the full potential of the computerized factory only if they provided workers with wide access to the system's potential, because the computer made possible the synthesis of hands-on knowledge with management.

Managers "found the new data-rich environment to be a humbling experience" because they saw a new complexity to the production processes they had tried to control. One manager put it this way: "If you like to control, then it is frustrating. Data opens up the organization. Everyone has to be more humble, modest, open."[35] The textualization of work and the resulting transparency of operations and individual performance made every worker more aware of and responsible for the whole. Together, they shared responsibility for creating and interpreting the computerized data, undermining centralization and attempts to dominate from the top down.[36] By the end of the 1980s it had become clear that the "smart machine" that for 30 years had been expected to concentrate power and information instead could disperse both through the company.

To this point, the discussion has implicitly dealt with large companies. Yet smaller firms have always been an important part of advanced economies. By the 1970s, mastery of high-tech machinery made it possible for a company with as few as ten workers to innovate and compete at a high level. In Italy, "innumerable small firms [were] specializing in virtually every phase of the production of textiles, automatic machines, machine tools, automobiles, buses, and agricultural equipment."[37] This "high-tech cottage industry" had emerged from the ranks of skilled craftsmen, many of whom left larger companies to set up on their own. These small companies not only created new jobs but paid high wages. "The innovative capacity of this type of firm depends on its flexible use of technology; its close relations with other, similarly innovative firms in the same and adjacent sectors; and above all on the close collaboration of workers with different kinds of expertise. These firms practice boldly and spontaneously the fusion of conception and execution, abstract and practical knowledge. . . ."[38] During the 1980s and the 1990s, the dynamism of small companies, in contrast to large corporations, was widely recognized. Peter Drucker, one of the most influential writers on business management, noted that small firms are agile and move more quickly to adopt innovations than corporate mastodons, many of which decentralized to meet this challenge.[39] Furthermore, historians found that "high-tech cottage industry" was not a recent phenomenon. Notably, Philip Scranton showed that a dynamic group of smaller firms has long been present.[40] Rather than see industrialization as the tale of consolidation into ever larger firms, in which mass production inevitably leads to the greatest efficiencies and profits in massive corporations, Scranton showed that the same ingenuity, quick response time, and batch production found in 1970s Italy had existed in the Philadelphia of 1910.[41]

Feminists point out that work is unevenly shared and rewarded. A survey sponsored by the American Federation of Labor found that 71 percent of working women feared they might lose their jobs, and that 30 percent of working women entirely lacked health insurance, prescription drug coverage, and pensions.[42] Such inequities have long persisted. In early-fourteenth-century Paris, women's wages in 130 occupations—including masons, shoemakers, smiths, money changers, mint workers, and candlemakers—averaged only two-thirds of what men received.[43] Precisely the same unequal wage ratio existed in the United States in 1999.[44] Such inequities have persisted despite repeated demonstrations that women can perform at the same level as men. For example, during both world wars British and American women showed that they could do factory work as well as men. A 1943 War Department guide for American managers declared: "Women can do any job you've got—but remember 'a woman is not a man.' A woman is a substitute—like plastic instead of metal."[45] Because of such denigrating assumptions, after each conflict, companies again marginalized women. Part of the difficulty also was that in the twentieth century women sought entry to a blue-collar workforce that (except in wartime) was shrinking. Making matters worse, the gulf between white-collar and blue-collar work has been widening, so that women who entered factories seldom found there a ladder to advancement. Often, they found only new rungs at the bottom. Women have arguably had more success in advancing through the hierarchies of the legal and medical professions and white-collar work. However, they seldom reached top management positions.

The problem lies in how machines and skills are embedded in cultural systems, which have made it hard for women, until quite recently, to become chemists, engineers, computer programmers, physicians, or executives. Cultural predispositions still define many jobs in terms of gender. Women have been subalterns— assistants to the males who control hospitals (nurses), corporations (secretaries), and dental practices (hygienists). Likewise, women have most of the lowest-paid jobs, including child care, cleaning, and working in restaurants.[46] Employment segregation reflects cultural values. Indeed, to the extent that machines now do the heavy lifting, divisions of work based on size, height, and strength (qualities that have often functioned as code words for gender) no longer apply. A woman can drive a forklift or manage an electric crane as well as a man. Technologies do not in and of themselves cause gender inequality. Rather, they can be socially constructed to restrict or improve women's access to some jobs. For example, housework has increasingly become women's work. Numerous studies have shown that during the twentieth century the hours of domestic labor did not diminish despite the adoption of electric vacuum cleaners, stoves, clothes washers, and dishwashers. A man can run a vacuum cleaner or a dishwasher as well as a women. But as such appliances entered the home, men and children tended to withdraw from domestic work, leaving mothers to do most of it alone, even as standards rose for child care, cuisine, and cleanliness.[47] This persistent inequality cannot be blamed on household appliances, but strongly suggests that technologies are socially shaped to perpetuate pre-existing cultural values and male privilege.

Men long dominated the engineering professions, almost to the exclusion of women. This has changed. By 2002, nearly half of the students at MIT, and 62 percent of the chemical engineering majors, were women.[48] Taking a long view, technologically developed societies clearly can choose to give women equal access to every sort of employment. Yet it cannot be assumed that this potential change will actually occur. In Britain, a higher percentage of women worked outside the home in 1800 than in 1900, which strongly suggests that there is no automatic correlation between advancing industrialization and more women workers or more equality for women.[49]

Rikfin's book *The End of Work*[50] was at best premature in predicting the inevitable disappearance of jobs in the global labor market. If assembly lines, computers, robots, the Internet, and automated telecommunications displaced millions of workers, during the 1990s growing consumer demand created even more jobs in exchange. If large corporations were often struggling, nimble small firms proved innovative, creating more jobs than the large companies lost. Even more surprisingly, average work time, after decreasing from more than 60 hours a week 1870 to about 40 in 1970, began to increase. In the 1990s, American laborers annually worked the equivalent of a month longer than in 1970, yet felt less secure in their jobs. "Rather than hire new people, and pay the extra benefits they would entail, many firms have just demanded more from their existing workforces. They have sped up the pace of work and lengthened time on the job."[51] In 2004 many German employers successfully negotiated an increase from 36 to 40 normal working hours per week, with no increase in pay. Similarly, in 2005 the French government gave up on the 35-hour work week. Contrary to expectations, the question of the new millennium is not "How shall we deal with either increasing leisure or technological unemployment?" Instead it is "Why has work become more demanding?"

One explanation for why people work harder is that they are being exploited. For 200 years, unequal distribution of the profits from more efficient production has remained an intractable problem. Early mill owners usually became rich, while mill hands barely subsisted. English industrialization, with its grimy cities and sharp class divisions, was later replicated in the Pennsylvanian coal towns of 1890, in the Chicago stockyards of 1900, in the North Carolina mill towns of 1930, and in the Asian and Latin American sweatshops of 2005. Income distribution did not march automatically toward greater equality; it was achieved only in some nations, usually after unionization and political conflict. Nor is the long-term trend necessarily toward more wage equality. Between 1970 and 1995, the richest 20 percent of Americans increased their slice of the economic pie from 40.9 to 46.9 percent, while the remaining 80 percent shared the loss.

These general statistics suggest that if computers and other technological innovations do not lead to mass unemployment, they may be used to increase income inequality. One obvious way is by "outsourcing" work to less expensive labor markets. In 2004 Scandinavian Airlines fired hundreds of office workers and outsourced their jobs to India, where wages are less than 20 percent of what they are in Scandinavia. British Airways had already sent 2,400 office jobs to India. Likewise, American Express, J. P. Morgan

Chase, and Standard Chartered Bank sent 13,000 jobs to India. Some European hospitals even began to send x-rays to Asia for analysis.[52] In the new millennium, white-collar work was moving from Europe and North America to emerging economies, creating insecurity among the remaining workers with routine jobs. Globalized work appears more promising to nations that import jobs. Their economies have boomed, and not all the jobs are menial. In Bangalore, India, by 2004 there were 150,000 engineers—more than in Silicon Valley. In contrast, American computer engineers have begun to experience unemployment, and it has become somewhat difficult to attract students into the field. Furthermore, because many innovations arise on the factory floor, outsourcing threatens to undermine Western economies, because they may innovate more slowly than their Asian suppliers.

As factory and white-collar jobs exit Western economies, new low-wage jobs seem to increase. In the ever more rationalized meat-packing industry, a manager boasted "We've tried to take the skill out of every step."[53] That made it easier to hire mostly unskilled immigrants. These unorganized workers earn one-third less than meat packers did in the 1960s, and they receive no health benefits until after six months and no vacations until after a year. With annual turnover of 80 percent or more, in practice most of these workers have no benefits. Moving slightly up the job ladder, new employees in a fast-food restaurant, an insurance company, or a telemarketing company are commonly confronted with extreme routinization, as their work and interpersonal behavior are scripted. In many such organizations, "the routines were based on a system and rules developed by a charismatic founder who was held up as a model for employees and managers." Management believes that its routines "codified the best methods"[54] and therefore reward conformity, not innovation. When the work is part-time and predictable, a company often makes only mild efforts to create a common mental outlook. But when work is full-time and customers must be courted (as in insurance, for example), companies are more likely to try to instill a "positive mental attitude" and to change how employees "approached life, viewed themselves, and thought about their experiences."[55]

Many new jobs demand few skills and little local knowledge. In the fast-disappearing neighborhood store, customers knew the proprietor, and shopping might include some small talk and gossip. In contrast, an instructor at McDonald's "Hamburger University" declared: "We want to treat each customer as an individual, in sixty seconds or less."[56] From the corporate point of view, workers are interchangeable and need not know the community. Chains such as McDonald's and Wal-Mart make little long-term commitment to their employees, and they pay considerably lower wages than the restaurants and smaller stores they replace. Wal-Mart, now America's largest retailer, has bankrupted the shops in many towns, leaving empty stores on Main Streets across the country. Its workers are not unionized. An audit of 25,000 of its employees found 60,000 cases of workers missing breaks mandated by law, and 15,000 instances where they worked through mealtimes. Wal-Mart's practices pressure other retailers to follow suit, notably in California, where the three largest grocery chains proposed to slash wages, health benefits, and pension plans. These reductions prompted a 138-day strike that involved 59,000 workers. They ultimately only managed to save their own skins, while wages and benefits for newly hired workers dropped by $8,000 a year. A baker who had just two more years until retirement, concluded: "With this kind of contract, the middle class is just becoming the lower class. But I'll vote for it. I want to go back to work."[57] Wal-Mart plans hundreds of additional stores, where it will continue to pay workers poorly, a trend that is not the fault of any technology per se. Rather, it emerges in the context of a society where unions are weak and minimum wages are one-third or more lower that in Britain, France, Germany, Holland, and other advanced Western economies.[58] Technologies are embedded in cultural systems, and wages are part of these systems. Wal-Mart not only expresses the general Western preference for efficiency in production over other values; it also expresses an American preference to pass on savings in efficiency to consumers and stockholders but not to workers. In making this choice, Wal-Mart also forces suppliers to adopt and embody the same values. When it demands lower wholesale prices in exchange for huge orders, suppliers must either press their workers to be more productive or pay them less. Wal-Mart embodies not only the economies of scale possible in mass distribution, but also American resistance to high minimum wages, unionized labor, and the welfare state. Thus, in the "new economy" many workers have longer hours because their jobs feel insecure, because wages are lower, and because employers have discovered that it is cheaper to pay overtime than it is to hire, train, and provide pension and health benefits to new employees.[59]

A second group, however, wants the longer work week because it gives them more money for consumer goods. Early in the nineteenth century, the average person had a modest wardrobe, often no more than one set of clothing for daily wear and a Sunday suit. The Western consumer today has a closet stuffed with clothing, some little used. The European or American also wants more space to live in than ever before, including not only a larger house or apartment, but ideally a summer home as well. In the 1920s, one car was enough for each American family. By the 1960s, each adult "needed" a car. Today, every person in the United States over 16 "needs" a car. Overall, it costs the average American family over $7,000 a year to own, insure, and operate automobiles, more than it spends on food.[60] Demand for more goods of all kinds has risen so fast that even when real wages increase, savings may fall. During the boom of the 1990s, the median American family had less than $10,000 in assets.[61] Middle-class Americans worked longer hours in order to possess more things, not least new laptop computers, mobile phones, automobiles, and other technologies that embody high mobility, interactivity, and success. Most of these purchases lose value each year and become worthless within 10 years. Continually replacing and upgrading them has become an incentive for overtime.

And there is a third group, made up of what Richard Reich calls "symbolic analysts"—people "who solve, identify, and broker new problems."[62] In the borderless world economy, they are in great demand. Their work is not routine but varied and interesting, which explains why many of them are "workaholics." They put in longer hours not due to economic necessity and not (only) because of an insatiable desire for more consumer goods. These are usually well-educated professionals who love their jobs, and even when away from the office never really leave it behind. Their mobile phone is always at hand; their portable computer is always on. Their identity has become so entwined with the job that family and friends often take a lesser place in their lives. Corporations have found that such employees will work especially long hours if food and coffee are freely available, and if they provide other amenities such as small kitchens, exercise rooms, and jogging paths. At Microsoft, refrigerators are perpetually stocked with free soft drinks and sandwiches. Far from seeking a life of leisure, the most highly educated often embrace an almost ascetic routine of long work hours punctuated by physical workouts. Such habits are widespread in Silicon Valley and in other high-tech communities.[63] Technologies were often introduced in order to save labor. Yet the more skilled people are, the more likely it is that they will use new machines and more efficient processes to increase their workload. Those dealing with advanced technologies seem particularly prone to seek new tasks, to work longer hours, and to consume more goods.

The pattern traced here suggests that technology probably will not be used in the way that H. G. Wells feared it might, to empower a highly educated technical elite that wields hegemonic control over the rest of society. Rather, use of advanced technologies in the workplace has eliminated some routine jobs but created others. It has increased the number of skilled workers and increased the pressure on all to perform. It has weakened management's monopoly on information inside the factory, changed the role of some labor unions from confrontation to partnership with management, and enabled many small companies to compete effectively or to find profitable niches. Yet nations allocate the benefits from efficient production quite differently. In the United States, minimum wages remain low while benefits and pensions have been eroded. In contrast, the wealthiest European nations pay double the American minimum wage and provide citizens with free health care. These differences are not rooted in technologies, but in cultural values.

Nations are hardly closed systems. They must compete in a global marketplace that puts cultural values under pressure. As corporations shift routine work to cheaper labor in Asia and Eastern Europe, it seems that Western workers can maintain their higher wages and benefits only by acquiring more education and embracing ever more advanced technologies. If there is a limit to these continual processes of work redistribution and retraining, it is not yet in sight.

How should a society select technologies? Can such decisions be left largely to market forces, or are corporations now so dominant that countervailing governmental powers are needed? What is the proper balance between the market and regulation? Consider three examples: (1) A Levittown homeowner who wants to build a new wing needs planning permission; once completed the work must pass inspection. In such cases, both design and workmanship are subject to government controls. (2) An airline worker has paid Sausalito-based Genetic Savings and Clone $50,000 for a kitten copy of her dead cat.[1] In this case, the government plays a lesser role. Why is there less control over the cloning of cats than over the building of homes? (3) New drugs require government testing and approval before pharmaceutical companies can sell them. When Congress recently relaxed the standards in order to speed the approval process, the pain relievers Vioxx, Bextra, and Celebrex went too rapidly to market and killed many patients (quite possibly more than 1,000) before being withdrawn.[2] Clearly, with drugs tight regulation and government testing seem necessary. These three examples show that while some areas have long been regulated (e.g. home building) or even tightly controlled (drugs), the only controls over some products and services are protections against fraud, e.g. supplying a kitten that is just a lookalike and not a true clone. But is cloning itself desirable? Should voters and consumers have a say?

In his late work "The Laws," Plato argued that a well-regulated state should scrutinize innovations before permitting their introduction.[3] He feared that new practices, even in children's games, could disrupt the body politic. Such caution long persisted, and even during the Enlightenment some remained suspicious of technology. But today Western societies leave acceptance or rejection of new devices largely to market forces. The characteristic liberal view since the eighteenth century has been that technological advances are the mark of a democratic people, and that only a nation free of feudalism, guilds, royal monopolies and other artificial restraints can make full use of individual initiative and ingenuity. The hallmark of such a society is a patent system that encourages innovations by protecting the rights of an inventor for a period of years. The American founding fathers valued patent protections and guaranteed them in the Constitution (article I, section 8). They passed the first patent law in 1790, which obligated the Secretary of State to evaluate applications and grant patents. The work proved so time consuming, however, that a second law in 1793 granted a patent to virtually all who applied, and left investigation of originality to the courts. Not until 1836 did the United States establish a Patent Bureau. It actively vetted claims and established exclusive rights to new inventions.[4]

On July 4, 1830, the famous orator Edward Everett declaimed on the patent system and democracy at an entirely appropriate location, the new mill town of Lowell, Massachusetts: "It is the spirit of a free country which animates and gives energy to its labor, which puts the mass in action, gives it motive and intensity, makes it inventive, sends it off in new directions, subdues to its command all the powers of nature, and enlists in its service an army of machines, that do all but think and talk." The proof of this argument was not theoretical but practical: "Compare a hand loom with a power loom; a barge, poled up against the current of a river, with a steamer breasting its force. The difference is not greater between them than between the efficiency of labor under a free or despotic government; in an independent state or a colony."[5] It therefore seemed no accident that industry thrived in England and the United States while languishing under despots. Closer to home, this argument explained why the slave South lagged behind the free-labor North. This laissez-faire view may seem reasonable in the case of inventing and marketing the mechanical reaper, more efficient stoves, or an improved screwdriver. But is there any limit? What if a corporation wants to sell nanorobotic red blood cells that can carry ten times as much oxygen as normal red blood cells? Such artificial cells might save the lives of people suffering a heart attack, but they might also give superhuman endurance to soldiers, athletes and criminals. What unforeseen effects could they have?[6]

The patent system had been designed to encourage innovation, but corporations discovered they could also use it to dominate markets. During the twentieth century, as research laboratories largely replaced the private inventor, technologies were selected less by voters or consumers than by managers and investors. Early textile mills and other factories had retained a few clever mechanics to build and repair machinery. Occasionally they invented an improvement. Innovations until at least the middle of the nineteenth century mostly came from private individuals, a few of whom, notably Thomas Edison and Arthur D. Little, established permanent research laboratories. As corporations grew, however, they overcame rivals by gaining exclusive access to patents. To ensure a continuous flow of innovations, in 1900 General Electric inaugurated the first corporate research laboratory. The inventor was now on salary and the firm owned any innovations created.

Corporate managers studied markets to find out where their best opportunities lay and encouraged their new research labs to tackle problems of immediate value. The GE staff could in theory do whatever they wanted, but they were strongly encouraged to improve the filament of the light bulb, to design high-tension lines that could withstand lightning strikes, and to improve electrical appliances, such as stoves and refrigerators. GE researchers did all these things and more, and executives began to realize that a corporation that kept acquiring patented improvements could retain market dominance almost indefinitely. Little wonder that every major corporation today has a research and development division, and that most inventors now work for them. One should not exaggerate the rapidity of this development, however. In 1921 fewer than 3,000 were employed in American corporate R&D, and even in 1946 the number had only reached 46,000. However, by 1962 the United States alone had about 300,000 doing corporate research, and their numbers mushroomed to more than 1.2 million in 2002.[7] In the same year R&D employed over 600,000 in Japan, 260,000 in Germany, and 157,000 in Britain, while the OECD nations as a whole employed more than 3.3 million researchers, collectively accelerating the pace of change.[8] There are still private inventors, but the advantage clearly belongs to corporate labs.

In 1972 the US Congress, having noted this dominance, authorized the Office of Technology Assessment (OTA). Unlike the older Congressional Research Service (CRS), which handled all sorts of inquiries, the OTA developed in-depth technical expertise, as it examined such topics as acid rain, the de-commissioning of nuclear power plants, missile defense, and how to cut energy use in homes and businesses by one-third, using commercially available technologies.[9] Congressional staff appreciated these reports. On occasion they were cited in hearings and bills, and OTA staff testified as often as 55 times a year. For example, in 1995 Congress decided to cease funding the Advanced Liquid Metal Reactor based on an OTA report. The agency more than paid for itself during the early 1990s, as it identified ways to save more than $400 million.[10] However, OTA was hardly a cheerleader for development. Compared to the CRS, whose reports were more factual than analytical, and typically took the form of a pithy 10–20-page summary that treated all options as having equal validity, OTA wrote more and drew conclusions. In doing so, it undermined the pet projects of some representatives.[11]

When OTA was formed, the political scientist Mulford Sibley argued that "technology is so vital a part of life, and collective life particularly, that we cannot afford to leave the fate of man to the hazards of the 'market' and accident." A market dominated by corporations is far different from what Adam Smith had described in 1776. Too often, "the most important matters affecting the future of mankind are not subject to public deliberation."[12] While legislatures vote on such matters as social security, new roads, and education, often they pay scant attention to major technological changes and their long-term implications. Instead, vital issues are often left largely in the hands of private enterprises, which decide what new machines and processes to pursue.

Governments are increasingly likely to intervene in the market not through prohibitions or regulations but by introducing subsidies or tax incentives to stimulate research in the private sector. Governments also fund grant agencies such as the National

Science Foundation, not due to a thirst for pure knowledge, but in hopes of improving the economy. Leaders frequently assert the need for new knowledge and innovation to guarantee future success. For example, in a speech before the Royal Society in 2002, Prime Minister Tony Blair declared: "Britain can be as much of a powerhouse of innovation—and its spinoffs—in the 21st century as we were in the 19th and early 20th century. The benefits in industry, jobs of quality, healthcare, education, and the environment can transform our future."[13]

Blair also realized that an increasing number of citizens resist corporate experimentation. He acknowledged that environmental activists had destroyed fields planted with genetically modified crops and that a planned laboratory at Cambridge that would use "primates to test potential cures for diseases like Alzheimer's and Parkinson's" was under threat because some animal-rights activists were ready to take violent action to prevent its construction. Therefore, Blair spoke of the need to produce "a confident relationship between scientists and the public" and of the need to "exercise the care and judgement to make scientific discovery a liberating, civilising force not a leap into the unknown."[14]

National leaders are caught between scientists eager to innovate and activists who feel cut off from the technology policy process and instead turn to the streets. During the 1950s and the 1960s, technology activists worked primarily within the political system. However, they often had little influence on decisions, as corporate and political leaders worked together with teams of "experts" to define and resolve issues before they formally reached lawmakers. All too often, hearings and deliberations could become an endorsement rather than a debate. For example, in the 1950s corporate lobbyists and other pro-nuclear "experts" convinced the US Congress that atomic power stations were the wave of the future and would produce very cheap electricity. Congress voted in favor, but with hindsight the decision was flawed. Since the late 1940s over 95 percent of federal support for alternative energies has gone to nuclear power, ignoring solar and wind power. This skewed the market toward what turned out to be an expensive atomic reactor system. At the time, however, "experts" promised to produce infinite fresh water supplies, eliminate oil dependency, banish air pollution, lower electricity costs, manufacture cheap fertilizer, and enable mankind to live comfortably under the sea, in the jungle, or even in outer space.[15] Subsequent experience showed that cheap nuclear power was a fantasy, as electricity from reactors costs more than that from windmills or gas turbines.

With a longer period of development, better nuclear plants might have emerged. By the 1990s it was clear that these early atomic reactors were uneconomical, and many closed down. However, the radioactive waste they had produced still must be stored, from 120 years (tritium), to 280 years (strontium 90) to 700,000 years (nickle 59), or even 16 million years (iodine 129).[16] Furthermore, the facilities themselves are so badly contaminated that they need to be fenced and guarded for generations. In contrast, Norway rejected atomic energy and today produces 99 percent of its electricity from hydroelectric plants. Denmark also rejected atomic energy and became world leaders in wind power. In 2005 more than 12 percent of Danish electricity came from windmills.

If relying primarily on experts proved a financial fiasco in the case of early atomic power, traditional American politics could also produce poor decisions. Dealmaking between interest groups, not analysis, led to the flawed interstate highway program adopted in 1956. Neither the Clay Commission that prepared the government report nor the Congress that embraced it distinguished between travel inside cities and travel between cities. Debate scarcely considered mass transit, but focused on whether to fund the highways through tolls, bonds, outright federal subsidies, or the solution eventually adopted, a gasoline tax creating a fund to build and maintain the 41,000-mile interstate system.[17] It subsidized 90 percent of a highway's cost but gave nothing to mass transit. This choice doomed most remaining streetcar systems to extinction, ripped apart the urban fabric, encouraged suburban sprawl, and gave truckers a competitive advantage over the railroads.[18]

These examples suggest that at times the market can be more reliable in selecting technology than the political process. Had Congress trusted in the free market and avoided subsidies, more cities would have built or maintained mass transit, fewer power companies would have built nuclear reactors, and the United States today would consume far less energy. Unfortunately, the American government has signally failed to create a coherent energy policy.[19] Today, nuclear power is being touted as the solution to the problem of global warming, and many developing countries see it as a way to escape dependence on fossil fuels. China has determined to build nuclear reactors instead of more coal-burning power plants. But such a "technological fix" may create at least as many problems as it solves.

Advocates of a "technological fix" argue that "motivating or forcing people to behave more rationally" is a "frustrating business." It is easier to develop a "technological fix" that accepts "man's intrinsic shortcomings and circumvents them. . . . One does not wait around trying to change people's minds: if people want more water, one gets them more water rather than requiring them to reduce their use of water; if people insist on driving autos while they are drunk, one provides safer autos that prevent injuries even after a severe accident."[20] Such "expertise" short-circuits public debate and democratic decision making. Furthermore, these were bad choices. Nuclear-powered desalination was a favorite technological fix advocated in the 1950s and the 1960s, but it proved uneconomical. A better approach is to recognize that rainfall limits the water supply and to find ways to work within these limits.

As for the idea that the best way to deal with drunk driving is to build safer cars, Scandinavian countries have shown that intensive public education and stiff penalties can deter drunk driving itself. The taboo was strong enough in Denmark to force a powerful politician to give up leadership of a major party and leave Parliament, his career in ruins, because he wrecked his car while driving under the influence. He hurt no one, but the public regarded the accident as proof of a fatal lack of judgment. Statistics show that drunken driving is far less a problem in Scandinavia than in many other areas. Moderate to high levels of alcohol are found in only 1 percent of Scandinavian drivers in evening leisure hours, compared to between 5 percent and 10 percent in the United States, France, or Canada.[21] In short, legal sanctions and social norms rather than a technological fix are at times the best way to curb the misuse of a technology.

Yet sometimes the technology itself needs fixing, because the market has functioned poorly, as was the case with the American automobile industry in the 1950s and the 1960s. In *Unsafe at Any Speed,* Ralph Nader documented the dangerously poor quality of some American cars, leading to Congressional hearings and the National Traffic and Motor Vehicle Safety Act of 1966. Nader later organized watchdog consumer groups who took on a variety of other technological issues, including nuclear power and renewable energy, food and drug safety, and pollution. He founded public interest groups that served as a countervailing power to government inefficiency and occasional cronyism with corporations. More recently, his Global Trade Watch has become a leading anti-globalization group, which tracks the International Monetary Fund, the World Bank, and similar organizations. Such grassroots activists focus on environmental protection, genetically modified foods, public health, the elimination of poverty, and technology policy.

By 1999 many activists no longer believed that political lobbying and lawsuits could rein in corporate research and development. At the turn of the millennium a revolt emerged, as thousands protested against the World Trade Organization and blocked the streets when it met in Seattle, Gothenburg, and Genoa. The protesters were a loosely organized coalition that included feminists, indigenous peoples, the Sierra Club, labor unions, organic farmers, opponents of genetic engineering, advocates of biodiversity, and anti-capitalists. Naomi Klein's *No Logo* inspired many in the movement by documenting how brand-name corporations exploited sweatshop labor in Asia and Latin America to eliminate jobs in North America and Europe.[22] Many protesters felt that the World Bank, the International Monetary Fund, and other international agencies dictated the form of growth and the technologies used to achieve it. Furthermore, during the 1970s and the 1980s the World Bank had an extremely conventional approach to energy in the developing nations, and took no interest in solar energy, even though many public and private organizations had found that it was cost-competitive for some uses, especially in deserts and remote areas.[23] The vast majority of the Seattle protesters were non-violent. While some media rather freely called the Seattle demonstration a riot, the tense situation owed much to police incompetence, as they failed to stop vandals looting stores, often failed to protect the delegates to the meeting, and fired gas and rubber bullets at peaceful pickets.

The mainstream media demonized the protesters and simplified their arguments into the slogan "anti-trade," but the opposition to the WTO was far more complex and interesting than that. The media might have analyzed the 43 essays in *The Case against the Global Economy,* a book published by the Sierra Club.[24] One of the contributors was the novelist and critic Wendell Berry, who continues to farm using low-tech methods. He declared that the most important difference between "the industrial and the agrarian, the global and the local" was "knowledge. The global economy institutionalizes a global ignorance, in which producers and consumers cannot know or care about one another, and in which the histories of all products will be lost."[25] As a result, consumers unwittingly participate in the destruction of distant ecological systems and the exploitation of workers elsewhere. Berry called for "a revolt of local small producers and local consumers against the global industrialisation of the corporations."[26]

How might the activism that worried Tony Blair and that disturbed meetings of the WTO be reincorporated into the political process? Rick Sclove has argued in *Democracy and Technology* that when new technologies are being adopted ordinary citizens typically play too small a role, often only after the most important decisions have been made.[27] When approving new products, governments that listen mostly to experts may fail to consider cultural and political effects. In contrast, the Dutch and the Danes have developed forums of representative ordinary citizens who interview "experts" and then formulate advice on technological policy. Ideally, every society should give citizens such an opportunity to influence the construction of technological systems. In the future, citizens are likely to demand more transparency and debate in technological decision making. Should corporations manipulate DNA to develop new life forms, use advanced robotics to build increasingly intelligent machines, or use nanotechnology to build substitutes for human cells? Are such technologies too valuable or too dangerous to be left largely under private control?

The relationship between technology and human freedom is more than a question of consumer choice, and more than a question of whether successive forms of mass communication improve or undermine political discussion. Beyond these important matters lies the fundamental question of who controls technological change itself. Leaving "the market" in control permits corporations with little hindrance or discussion to disseminate thousands of products that foster lasting changes in everyday life. When corporations introduce genetically modified life forms, make goods that are not biodegradable, or sell products that are difficult to recycle, voters clearly need to become involved in technological policy. Steven Goldman has argued that no innovations should be introduced without representation: "Science and technology policies have a social impact comparable to that of taxation policy in the colonial period. In 1776, political freedom entailed the right to a voice in taxation decisions because these decisions were primary forces in shaping the fabric of personal and social life."[28] Just as the colonists demanded a voice in tax policy, today citizens need meaningful representation when technological decisions are made.

Politics can focus too exclusively on social programs and national security. The citizen feels secure when the police have arrested terrorists who avowedly threaten their way of life. But, as Sibley noted, "one never hears of the FBI rounding up those who introduce new machines which are depopulating the country-side, depriving men of their vocations of a lifetime, destroying much of the earth, polluting the atmosphere," or otherwise transforming social and political life. "Until Americans [or other nations] develop the law, standards, practice, and organization for deliberate public introduction or rejection of complex technology, their supposed self-government will remain largely a pretence."[29] The best uses of technology can only emerge when consumers recognize that new machines are not inevitable and that their uses are not ordained. Nanotechnology does not have to be used to make new kinds of red blood cells. "Cognition enhancing drugs" may be possible, but if so should they be widely used? Which aspects of human intelligence ought to be enhanced and which not?[30] Voters, not "the market," should decide such issues.

Likewise, research on aging has such enormous political implications that decision making cannot be left to corporate boardrooms. Even the free-market conservative thinker Francis Fukuyama, who approves marketing genetically modified plants, is not ready to hand human experimentation over to the private sector.[31] Some scientists, including Aubrey de Grey at Cambridge, are working toward a future when people will age very slowly, if at all. De Grey's website (www.gen.cam.ac.uk) declares that his work "is an engineering project, in the same way that medicine is a branch of engineering" and goes on to assert that "aging is best viewed as a set of progressive changes in body composition at the molecular and cellular level, caused as side-effects of essential metabolic processes. These changes are therefore best thought of as an accumulation of 'damage', which becomes pathogenic above a certain threshold of abundance. . . . [the] strategy is not to interfere with metabolism per se, but to repair or obviate the accumulating damage and thereby indefinitely postpone the age at which it reaches pathogenic levels." The researchers expect that when cells wear out, cloned and cultured stem cells can be used to replace them, or if cells are diseased or "unwanted" they can be eliminated through a variety of techniques. Some hope to extend life 50 years or more by 2100, and expect eventually an average life expectancy of 1,000 and perhaps even 5,000 years.

If this proves possible, is it desirable? The chair of President Bush's Council on Bioethics thinks not. He rejects such advances as "unnatural" and argues that the certainty of death gives life definition and meaning.[32] In contrast, a leading researcher remarked to me over dinner that ours might be one of the last generations that will have to die. What would extremely long life expectancy or near immortality mean to the political and social system? Would society become a giant nursing home for an ancient population, or would vigorous productive years stretch out for centuries? Would world population grow to completely unsustainable levels? Would anti-aging technologies be equally available to all, regardless of wealth, talent, or education? What would politics be like if the average voter were over 100 years old? Would retirement come at 150, 300, or never? Furthermore, if machines are made intelligent and people are engineered, the difference between them may become blurred. In such a cyborg world, what constitutes a person? Who has human rights? Who has voting rights?[33] Markets alone cannot deal with these deeply political questions. However, before citizens can think well about such matters they need information. Neither newspapers, nor television, nor radio networks conduct regular debate on technological policy, however. In theory, the media facilitate open debate and the exchange of ideas, but in dealing with technology most cultures do this job only haphazardly.

In the fifteenth century the new technology of printing was first used to produce Bibles. Politics itself was not openly discussed in print. In the sixteenth century, governments began to publish proclamations and explain regulations, and printers issued accounts of notable events. By the early years of the seventeenth century, weekly accounts of foreign news appeared in several European countries. Governments were suspicious of these summaries, however, and sometimes suppressed them. How freely was the technology to be used? In 1644, John Milton addressed an essay in defense of "the Liberty of Unlicenc'd printing" to the English Parliament, advocating the then-novel idea of a free press.[34] The US Constitution later included freedom of the press in the Bill of Rights (1790). Yet well into the nineteenth century, European governments regulated publications and newspapers through taxes on paper, licensing fees, and outright censorship. Subsequently, legislation shaped every form of communication, notably radio and television.

The rise of newspapers coincided with increasing political activism among their middle-class readers. The classic argument maintains that a nation fosters freedom "when the means of communication are dispersed, decentralized, and easily available," but authoritarianism is "likely when the means of communication are concentrated, monopolized, and scarce. . . ."[35] Europeans only accepted this correlation during the nineteenth century. A historian of journalism summarized this as follows: "In 1815 press freedom in Europe was an idea, a dream, an experiment, a fearful dread; by 1881 it had become an enduring institution, its most admired text enshrined in the French Press Law passed in July of that year."[36] Norway is a classic example. It had few newspapers in the eighteenth century, but by 1870 it had eighty of them, some voicing demands for independence from Sweden.

Yet it is hard to argue that introducing the printing press automatically created investigative reporting, political pluralism, and widespread literacy, all of which are needed to support democracy. The printing press was early introduced into the Ottoman Empire, which long restricted its use.[37] At first, Armenians and Jews could print works in their own languages, but printing in Arabic or Turkish was prohibited. The government relaxed these restrictions during the eighteenth century, but few newspapers emerged even in the nineteenth century. The Ottoman Empire provides little evidence for the "soft determinism" argument that printing leads to newspapers and then to democracy. An autocratic state that was both wealthy and powerful at the time of Guttenberg, the Ottoman Empire long resisted expansion of freedom of speech. Turkish newspapers emerged late, and they were long either published by or closely supervised by the State, which used the press to increase its power. Not until the twentieth century did Turkey develop communications with greater openness, and some states once part of the Ottoman Empire are still undemocratic.

As the example of the Ottoman Empire suggests, cultural contexts shape communications. Mass communication is not inherently hegemonic or democratic; its social construction can go either way. In the United States, freedom of the press emerged out of a pluralistic world served by many small printers, and the First Amendment to the Constitution was framed to protect them. Two centuries later, when Americans and Europeans sought to regulate mass-media networks, however, they erroneously assumed that only a few radio stations would be able to broadcast using a limited spectrum.[38] Most European countries opted to control this new technology through nationally owned stations; the United States licensed private enterprises and regulated them, in sharp contrast to how they treated newspapers. Anyone could start a newspaper and seek readers, but radio and television were defined as inherently oligopolistic markets. The limited broadcast spectrum was a legal construction that bore little relationship to technological possibilities. As the potential number of stations kept increasing, first with the development of FM radio and later with TV cable systems and satellite dishes, Americans and later Europeans opened media markets. Yet already established media corporations remained dominant, and today ten conglomerates control much of the world's media. Some are well known, such as Time Warner, Bertelsmann, and Sony. But how many are aware of the Liberty Media Corporation, which owns 20 television networks, including Discovery and USA Network, 14 TV stations, two production companies, six movie studios, 21 radio stations, and 101 magazines?[39] The power and international reach of such organizations makes them difficult to monitor or control. Permitting broadcasting to be a private oligopoly stifles democratic selection of new technologies, both by undermining democratic debate and by reducing the play of market forces. In theory, but not always in practice, governments should promote a more critical and competitive press. That they often do not is a reminder that politicians dislike being investigated or criticized. But it is extremely important that new technologies be investigated and criticized.

Each new form of communication, from the telegraph and telephone to radio, film, television and the Internet, has been heralded as the guarantor of free speech and the unfettered movement of ideas. They do not automatically function this way, however. It is probably true that democracy thrives when the press functions as a watchdog against government corruption or capture by special interests. Yet the media also legitimizes authority and sets the limits of acceptable public discussion. Nothing inherent in communication hardware ensures that it will be used to promote democracy or to conduct balanced discussions of new technologies.

Political manipulation is easier with centralization or monopoly of broadcasting. For 40 years, undemocratic communist regimes in Eastern Europe tightly controlled the press, restricted use of copying machines and typewriters, and resisted the introduction of personal computers, because all of these things gave the populace access to information. The fax machine and the photocopier were useful to Boris Yeltsin. During the August 1991 coup attempt, his announcements were photocopied and faxed around the USSR, providing alternative information at a time when the coup plotters controlled the state media. On the other hand, in 2005, the Chinese government still controlled television, radio, and newspapers, and it restricted access to Internet sites critical of the regime. Technologies such as printing, broadcasting, and the Internet help democracies thrive to the extent that people design and use them for the free exchange of ideas, but they are not automatically on the side of democracy.[40]

Some researchers find American television hegemonic, and argue that viewers merely seem to have choices as they select among essentially similar radio and TV stations. The more people watch, it seems, the less likely they are to attend a political meeting or to vote. Much programming, as Todd Gitlin observed, is designed to stir up intense emotions briefly, and viewers soon learn to jettison each temporary feeling and to zap around looking for more excitement. He argues that the long-term result is a demobilized electorate. Watching television, the "ceaseless quest for disposable feeling and pleasure hollows out public life altogether. If most people find processes, images and sounds more diverting, more absorbing, than civic life and self-government, what becomes of the everyday life of parties, interest groups, and movements, the debates, demands, and alliances that make democracy happen?"[41] The political scientist Robert Putnam made a similar argument in *Bowling Alone,* concluding that intensive use of the media undermines civic life. It seemed that the more people relied on television as the primary form of entertainment, the more they disengaged from political life.[42]

This disengagement is particularly notable in the area of technology policy. To a considerable degree, the media treat inventions as investment opportunities not as social and political questions. Most people unconsciously assume that technology "belongs" on the business pages not in the politics section of the newspaper. When controversy does arise, the media are likely to frame a question in narrow economic terms, such as "How to pay for a new airport?" and not "What are the environmental consequences of subsidizing airports and airplanes instead of high-speed trains?" or "What form of transportation is most desirable if environmental impacts are included in the analysis?" Likewise, the media discussion of nuclear power in the 1960s and the 1970s did not help the public to understand the technology as a series of choices. Rather, 1,200-megawatt light-water reactors modeled on those made for nuclear submarines were presented as the ideal or appropriate form of nuclear technology. Yet such reactors had been chosen by the military, working with General Electric and Westinghouse. They did not emerge from a nuanced public debate, based on a range of choices that engineers suggested.[43] Only very occasionally have the technological attributes of large projects become salient in public discussion, notably in the American decision not to build the SST airplane, which would have competed with the British-French Concorde and no doubt would have lost just as much money. More commonly, prestige projects such as the Apollo Space Program or the Strategic

Defense Initiative have been selected and promoted for political ends, without always selecting the optimal technological solutions.

By the 1970s, Jürgen Habermas had concluded that public opinion had become too thoroughly influenced by the mass media to allow for creative discussion. The "communicative network of a public made up of rationally debating private citizens has collapsed; the public opinion once emergent from it has partly decomposed into the informal opinions of private citizens without a public forum and partly become concentrated" into the expressions of institutions. The private person no longer operates in the public sphere known by Thomas Jefferson, but instead is "caught in the vortex of publicity that is staged for show or manipulation."[44] The increasing sophistication of "spin doctors" suggests that Habermas was correct.

When the Internet emerged in the decades after Habermas wrote, however, it seemed to offer a new forum for grassroots democracy, where individuals with similar concerns could find one another and form cyber-communities. Some of the New Left from the 1960s latched on to the new medium, particularly in the San Francisco area and at universities. Starting in 1979, they formed discussion groups, which mushroomed to 5,000 sites by 1987, and grew exponentially to reach 2.5 million users just five years later. The early Internet seemed a new public sphere democratically open to all, which might replace the one that Habermas believed television and radio had vitiated. In *The Virtual Community*, Howard Rheingold called these cyber enthusiasts "homesteaders on the electronic frontier."[45] He hoped that their collective discussion would enable groups of like-minded people to develop a "collective intelligence" that was greater than that of any individual members of the group.[46] This goal may seem rather utopian, but through the Internet every citizen does have the potential to communicate with a large number of other voters at almost no cost compared to buying advertising space or making a mass mailing. The amount of technological information easily available to those with web access is far greater than that in most public libraries.

In the 1990s, political candidates began using the Internet to present programs in greater detail than they could over the airwaves or in print. Government agencies and lobby groups also provided more information to citizens who looked for it. Yet the Internet does not automatically enrich public discussion, but can only do so as the result of conscious public choices. In the 1990s commercialization undermined hopes that the Internet would primarily function as a free space of public discussion. Instead, advertisements, solicitations, pornography, and fraud are rife. By 2005, more than half of all e-mails were unwanted spam. Similarly, television and radio when first introduced were expected to enhance democratic discussion. But when television is used to broadcast government meetings and hearings, few citizens watch unless they are dramatic events, such as the Watergate hearings or the Clinton impeachment proceedings. Likewise, just putting information on the Internet does not guarantee that anyone will find it, read it, understand it, and then take action.

The mainstream media can be so compelling that they define the limits of debate. Thus arises "repressive tolerance," in which all views can be permitted because most protests and many practical suggestions will be overlooked. In allowing expression of all views, the government shows it is not repressive As Herbert Marcuse put it, "the exercise of political rights (such as voting, letter-writing to the press, to Senators, etc., protest-demonstrations with a priori renunciation of counter-violence) in a society of total administration serves to strengthen this administration by testifying to the existence of democratic liberties which, in reality, have changed their content and lost their effectiveness."[47] Under a regime of repressive tolerance only a minority of the population follows the news closely, and dissidents, minorities and eccentrics have no impact if permitted to speak freely, since major newspapers and television networks ignore them. The Internet may yet illustrate this process, despite its democratic promise. Initially, it seemed to empower individuals and grassroots organizations. Yet the older media have successfully transferred their brand names to the Internet, where large newspapers and television networks continue to define and disseminate much of the news. Whether this new medium turns out to be an instrument of democracy or an example of repressive tolerance remains to be seen.

From the ancient amphitheater to the restrictions on printing to the distribution of broadcasting rights, governments have shaped the media. This is only one more example of how political and social constraints shape technologies, as was discussed in chapter 4. In practice, news and analysis of technology gets little airtime in most nations and is scarcely visible. The task of gathering and evaluating information therefore falls primarily to legislatures, where few elected members have science or engineering degrees. Indeed, in their ignorance legislatures at times want to rewrite natural laws. A representative in the US Congress once suggested repealing the second law of thermodynamics when he heard a scientist say that it stood in the way of his proposal. Indiana's state legislature once voted 67–0 to simplify the value of π to 3.2.[48] However easy this might make calculations, it would be a dangerous constant to use when building arched bridges. A study of how congressmen make technical decisions found, not surprisingly, that they do not sort out conflicting scientific testimony. Rather, one congressmen freely admitted, "you use the scientists' testimony as ammunition. The idea that a guy starts with a clean slate and weighs the evidence is absurd."[49] Like lawyers trying a case, each side marshals scientific testimony. They focus on winning, not on technical feasibility or scientific accuracy, and corporate lobbies are happy to help find obliging witnesses.

Given such practices, there was little demand for what the Office of Technology Assessment had to offer, and the Republican-dominated Congress de-funded and closed it in 1995. Can other agencies fill the void? The Congressional Reference Service cannot, as it has less technical competence, writes shorter reports, and works primarily in house. The General Accounting Office does after-the-fact assessments that focus on implementation and efficiency. An auditing and accounting department with little expertise in technology, it does retrospective evaluations not forward-looking policy recommendations. The Congressional Budget Office might seem a possibility, but its staff consists almost entirely of economists, who focus on the immediate effects of decisions already taken. On the positive side, if no agency provides OTA-like expertise, since 1973 the Science and Engineering Fellowship Program has reduced scientific illiteracy by placing experts on congressional staffs.[50] Unfortunately, each serves for only one year. They do not write independent reports but rather improve evaluation of information generated elsewhere. In short, while 3.3 million researchers around the globe innovate more rapidly than at any time in history, the US Congress has few analytic powers at its command. Rather, it must contract for special studies from outside institutions such as the National Academy of Science, an approach that is time-consuming and often impractical.

In contrast, European nations have developed this capacity, inspired by the now-defunct OTA. Britain and the European Union have adopted two inexpensive approaches. In Britain, a small office produces four-page balanced overviews on each technological issue before Parliament. Its small permanent staff writes them, after consulting outside experts. The European Parliament has established an agency that does no writing in house but contracts for reports. The Danish government took another approach. It set up an institution independent of Parliament, with a dual mandate, both to commission specialized studies and to involve the public in participatory assessment projects and "scenario workshops." These have been adapted and used elsewhere.[51]

Indeed, this "deliberative polling," has been tried 25 times in various countries, including the United Kingdom, the United States, Australia, Canada, and Denmark.[52] Just as the ancient Athenians at times selected a committee by lot to discuss a problem and suggest its solution, deliberative polling offers a representative sample of citizens a modest honorarium to spend several days at a comfortable conference hotel to discuss an important issue. They receive a packet of information to read in advance. Participants meet in small groups and develop questions for panels of experts. At the end of the process, participants have moved from a spectrum of fragmentary views to several focused alternatives. Unlike a trial jury, they are not required to reach unanimity. Rather, each participant completes a poll at the end of the weekend. The goal is to develop informed positions that represent the community's opinions. Deliberative polling is not merely experimental; it has been used to make decisions. In Texas, between 1996 and 1998 eight utility companies used deliberative polling to decide how to produce additional electricity.[53] To their surprise, after considering all the factors involved, customers wanted more renewable power and they were willing to pay a higher monthly rate in exchange for it. The utility companies added 1,000 megawatts in capacity, primarily as windmills.

But even if legislatures and other institutions (e.g., the World Trade Organization and the World Bank) have the advice of OTA-like institutions or adopt deliberative democracy, no one can foresee any device's full implications. Recall the unpredictability of technology, discussed in chapter 3. Five years after Edison invented the phonograph, neither he nor anyone else saw that its most important and lucrative use would be to record and play back music.[54] Those who created the early version of the Internet in the 1960s and the 1970s were unable to sell it to AT&T. No one foresaw the importance of e-mail, much less the World Wide Web. Why suppose robotics engineers or geneticists working to extend human life and intelligence will anticipate any better? Democratic deliberation about the desirability of new devices is essential, but invention is often the mother of the unforeseen. Neither legislators, nor experts, nor ordinary citizens can anticipate all the possible uses and effects of new machines. In the end, technologies without obvious defects will get a trial, but the best way to do this might be to combine deliberative polling with the sorts of controls used to evaluate new drugs. Society needs to take an incremental approach that cautiously tests new devices in the market, monitoring uses and effects, and reserving the power to recall. Vigilant citizens and government agencies will have to exercise constant oversight, for a society at the cutting edge of technological change confronts both opportunities and unforeseeable risks.

9 More Security, or Escalating Dangers?

We use tools and technical systems to protect ourselves, most obviously by constructing shelter. Sometimes houses collapse in an earthquake or a tornado, and occasionally a fire breaks out. Over time, we have learned how to make safer buildings. Until the late nineteenth century, entire cities occasionally burned down. That is far less likely today. For millennia, many improvements have made life more secure, protecting people from the elements, ensuring supplies of food and water, and widening the margin of safety. The scale of these efforts has grown far beyond the individual's capabilities, and whole systems of technologies have become specialized in building trades, housing inspectors, fire departments, and insurance claims adjusters. If one asks whether we have learned how to build better and safer buildings, the answer surely is Yes.

Over time, governments have increased their responsibilities for technologies of the infrastructure. Indeed, the modern state relies on engineering and architecture to demonstrate its legitimacy. Governments build roads, canals, bridges, public buildings, airports, harbor facilities, and military installations. So long as these function reasonably well, the state demonstrates its competence and by implication underscores its right to rule. If in addition these structures are beautiful, impressive, or even sublime, they may become important symbols, like the Roman Coliseum or the Great Wall of China. Public works that project national power are by no means a new phenomenon; witness the excellent Roman roads. Early modern France built many roads, grand fortresses, and canals not only to improve communications and defense, but also to demonstrate its legitimacy. Such "incremental acts of engineering can become politically significant because they address the fundamental problem of politics: creating commonality out of heterogeneity."[1] Technologies have long been fundamental to building the state, both as literal articulations of power and as legitimizing structures.

As engineering projects became emblematic of the state, however, technical failures had the potential to embarrass the government, question its competence, and force officials to resign in disgrace. Until the nineteenth century, people usually considered accidents and disasters, such as explosions, fires, collapsing buildings, or shipwrecks, to be inexplicable "acts of God." Often, people interpreted them as divine punishment. By c. 1900, however, they increasingly demanded scientific explanations. Geologists could show where, why, and to some extent when earthquakes were likely to occur. Engineers could explain how a bridge failed or what component of a steam engine was responsible for a burst boiler. The older moralizing tradition, which saw accidents as God's chastisement, disappeared, and citizens assessed risks, mistakes, and culpability. Formerly inscrutable events became legible to the safety engineer, the tort lawyer, the insurance agent, and the judge.

Accidents get maximum attention in the press, where "if it bleeds, it leads." Spectacular technological failures, such as runaway trains, airplane crashes, and burning skyscrapers, also recur as the themes of films and novels. Wolfgang Schivelbusch presciently observed that "the more efficient the technology, the more catastrophic its destruction when it collapses. There is an exact ratio between the level of the technology with which nature is controlled, and the degree of severity of its accidents."[2] The railroad usually is safer than most forms of travel. But when a speeding train goes off the track, the passenger has little chance to avoid serious harm. The greater the speed, power, and steam pressure, the more destructive a malfunction becomes. If an axle breaks on a farm wagon, it may cause minor injuries; when an axle on a railway engine broke outside Paris in 1842, many passengers died.

Steamboat explosions were spectacularly fatal because passengers were unavoidably near the boiler that gave way. When superheated water under high pressure escapes from a rupture, in less than a second it expands into scalding steam and blasts from one end of a steamboat to the other. No one has time to take cover or escape. A journalist reported what happened when all the boilers on the Mississippi steamboat *Clipper* blew up simultaneously in 1843:

. . . machinery, vast fragments of the boilers, huge beams of timber, furniture and human beings in every degree of mutilation, were alike shot up perpendicularly many hundreds of fathoms in the air. On reaching the greatest height, the various bodies diverged like the jets of a fountain, in all directions—falling to the earth, and upon the roofs of houses, in some instances as much as two hundred and fifty yards from the scene of destruction. The helpless victims were scalded, crushed, torn, mangled, and scattered in every possible direction—many into the river, some in the streets, some on the other side of the Bayou, nearly three hundred yards—some torn asunder by coming in contact with pickets and posts and others shot like cannon balls through the solid walls of houses at a great distance from the boat.[3]

Gruesome stories written in an equally graphic style became staples in the press, which reported railroad collisions, theatre fires, and bursting dams with somber enthusiasm. Some disasters captured headlines for days. The failure of a dam above Johnstown, Pennsylvania sent a 36-foot wall of water smashing through the city, obliterating buildings and even the pattern of streets, and killing more than 2,200 people.[4] In 1975, a much larger disaster occurred in China, where after heavy rainfall 62 dams collapsed in a chain reaction as upstream failures sent walls of water downstream to smash into dams. More than 26,000 people died.[5] In *Man-Made Disasters,* B. A. Turner views such events as the product of "cultural disruption" in a set of beliefs about hazards embodied in formal rules and procedures.[6] Turner developed the notion of a "disaster incubation period" when small inaccuracies or inadequacies become embedded in different parts of a system and accumulate without being noticed. No single mistake is usually the cause of a disaster, which rather comes about through an unforeseen interaction of several small miscalculations or errors. For example, Westinghouse manufactured a backup system for nuclear reactors designed to shut them down in an emergency. However, the backup systems used electrical signals that resembled those used in normal operation. At times, the two signals interacted, which simultaneously disabled both systems.[7]

Human intervention also can worsen "natural" problems. For example, in some places attempts to control malaria through extensive spraying with DDT initially eliminated much of the mosquito population, but ultimately "succeeded only long enough to produce DDT-resistant mosquitoes and human populations with little immunity [to malaria], leading in turn to intensified outbreaks."[8] Likewise, if efforts at fire control in national forests work for a time, this prevention allows an unusually thick layer of dead trees, brush, and leaves to accumulate on the forest floor, ready for conflagration. The same is often true in housing developments in forested areas.[9] The longer a fire is prevented, the more fuel piles up, leading to conflagrations far more intense than would normally occur. Similarly, constructing new levees can make floods more destructive, by preventing high water from spreading out and slowing down, instead channeling larger amounts downstream.[10] Averting inundation on the upper Mississippi increases the height of the flood in Arkansas and Louisiana. In each of these examples, human attempts to control nature have not solved a problem but intensified it. The scale of the disaster is often further escalated by other human errors— for example, building houses on river flood plains or along the Florida coast, where hurricanes regularly blast through.[11]

Many technical problems arise unintentionally, especially in rapidly developing industries where innovation creates unknown safety issues.[12] Asbestos once was widely adopted to make buildings fire-resistant, before anyone knew it caused lung cancer. In the 1970s several DC-10 airplanes crashed because their cargo doors blew open, causing rapid depressurization and buckling of the floor between the cargo hold and the cabin. When the floor broke apart, it damaged the hydraulic and electrical systems, making it impossible to control the plane.[13] Nor did anyone foresee that certain "inert" artificial chlorine compounds in millions of refrigeration coils eventually would escape and drift into the upper atmosphere, where they would begin to destroy the ozone layer. Nor could genetic engineers foresee that a plant they had designed to ward off insects might also release poison that kills valuable microorganisms in the earth, a poison that might even leech into the water supply.[14] To some extent, such dangers can be contained by tightening the statistical analysis of tests, by running more trials, and by including more people in evaluations of new products. Nevertheless, "as technology becomes more sophisticated in its manipulations of information, both biological and electronic, the possibilities for unexpected effects ramify beyond control."[15] Some disasters could have been avoided, notably the faulty cargo doors in the DC-10, a problem detected before the deadly accidents.[16] Likewise, some corporations tried to overlook the danger of asbestos in order to maintain profits, but later went bankrupt, due to lawsuits from poisoned workers.[17] But even rigorous precautions cannot always prevent accidents once a new product is released and used in conjunction with many other technologies.

As the risk of unforeseeable accidents increases, some consumers reject innovations not due to definite knowledge of risk, but rather because of a generalized fear of the unknown. For example, some consumers choose organic foods in order to minimize exposure to potential dangers that might arise from eating a cocktail of food additives, artificial colorings, traces of pesticides, the residue of chemical fertilizers, and genetically modified plants. Since no one knows what the interactions might be after ingesting random combinations of these things, avoiding them completely seems to many to be the most rational option.

Overall, the adoption of ever more advanced technologies leads to a paradoxical result. Individual machines are less dangerous. Steam boilers blow up much less frequently than they did 150 years ago, because engineers understand metal fatigue, steam pressure, and safety gauges. Airplanes and automobiles are safer than they were 50 years ago. However, the ensemble of technological systems in society has become so complex that it is difficult if not impossible to foresee how they will interact. The space shuttle *Challenger*, with thousands of well-designed components, disintegrated shortly after takeoff because its inexpensive O-rings did not function properly at low temperature.[18] In 2003, the electricity failed in the northeastern United States and much of eastern Canada because of a small malfunction in rural Ohio. In each case, the individual parts of a technological system interacted to create unforeseen problems.

Thus, if people have used technologies to increase their safety, they simultaneously risk unforeseeable accidents and even disasters that arise from the interplay of changing technical systems and new circumstances. The greater the power of systems, the more serious is potential failure. This conclusion applies with even greater force to the military, which faces more severe difficulties than railroad safety engineers or quality-control specialists. Technological failures in civilian life typically occur due to miscalculation or unanticipated circumstances. The military faces not only these problems, but also intentional attacks from enemies who probe for weak points.

Do weapons make people safer? Consider this question on a personal level. Does ownership of a handgun confer safety on a family? In 1998, 31,708 Americans died from gunshots. Of these, 17,424 were suicides. More than 1,000 of the deaths were due to accidents. Americans also used guns to murder 12,102 people, many of them in domestic disputes. Roughly two-thirds of all gun-related deaths were self-inflicted accidents, suicides, or murders within the circle of family and friends. Guns purchased as protection against criminals or intruders all too often killed the people they were expected to protect. Americans found guns ready to hand when they were careless, angry or depressed. In countries where guns are not so easily available, the homicide and accident rates are lower. Children in the United States were five times more likely to die from gunshots than in any of the other 25 most prosperous nations, including Japan, England and Wales, Sweden, Germany, Australia, and Italy.[19] A Harvard study confirmed the correlation: "In states with more guns, more children are dying."[20] However, one cannot simply scale up this example to understand the consequences of creating permanent armies of adults.

The technologies of war include weaponry, fortification, medicine, supplies, transport, and communication. Superior weapons can tip the balance in battle. Strong defenses—whether fixed installations such as walls, moats, and forts, or individual shields, chain-mail armor, or bulletproof vests—can help a small army defeat a larger foe. Better medical care and sanitation keeps more troops in the field. In the American Revolution 75 percent of those treated in hospitals died. In the Vietnam War this figure had dropped to only 13 percent.[21] Armies also need supplies. In the American Civil War (1861–1865) the factories of the North gave its troops decisive superiority over the South in provisions, arms, uniforms, and equipment. American factories proved just as important in the two world wars of the twentieth century. Likewise, better transportation ensures that vital supplies are available when needed. Hence the importance of railroads in the American Civil War, trucks in World War I, and cargo planes in World War II. Superiority in just one such area can be the difference between victory and defeat.

Consider communications as a final example. Armies long depended on the runner and the carrier pigeon. Indeed, as late as 1939 the US Army still used pigeons, because the radios in its tanks often did not work. (Eventually the Army adopted, with little modification, the FM system used by the Connecticut State

Police.[22]) French tanks had no radios at all in 1939, whereas German tanks had excellent radios. The French might have learned from the Japanese, who used radio effectively in their 1905 victory over the Russian army.

If technologies are at the center of warfare, however, the military tends to rely cautiously on what has worked in the past. The Roman armies held their empire with weapons the ancient Egyptians would have recognized, although there had been some improvements in metallurgy and design. Indeed, "from 3500 B.C. to 1300 A.D. [military] technology evolved very slowly."[23] The crossbow, found in China as early as 220 B.C. and known in the late Roman Empire, was not widespread in Europe until the medieval period. Even when a new weapon was adopted, older systems tended to coexist with it for some time. Pikes continued to be used for more than 100 years after firearms were introduced, partly because the early muskets and pistols were inaccurate, had a short range, and took time to reload, but also because the military remained a conservative institution. Late in the nineteenth century the US military rejected several working versions of the machine gun.

The Chinese invented gunpowder but did not use it in weapons, which illustrates once again that technologies are hardly deterministic. But when Europeans obtained gunpowder, during the Renaissance, they used it to change the scale and the nature of warfare. The era before canons, rifles, and pistols was not a golden age of chivalry. Yet gunpowder made a qualitative difference. A cannon firing grapeshot could annihilate distant ranks of soldiers. Cannonballs could destroy walls that had withstood catapults, and could tear down the sails of distant ships. With heavy ordnance, a navy could destroy a city without sending a single man ashore. In the Napoleonic Wars the British fleet bombarded Copenhagen with incendiary shells, burning much of it to the ground. Gunpowder made warfare nameless and faceless. Technical superiority often was crucial to victory. The British could not penetrate China until c. 1840, when they brought steam-powered gunboats onto Chinese rivers and forced their way to the Grand Canal. Beijing had little choice but to sue for peace. Later triumphs in Africa were also based on the use of gunboats.[24] It is not an incidental detail in Joseph Conrad's *Heart of Darkness* that the European trading company controls the river through steam navigation. Americans adapted steam power to inland navigation even more rapidly than the British, using paddlewheelers to gain military and economic control of the Ohio, Missouri, and Mississippi Rivers. If steamships provided access, breech-loading rifles, machine guns, and rapid-firing light artillery ensured dominance. With their superior firepower, Europeans and their native allies repeatedly defeated much larger forces. In 1899, in Chad, 320 French forces (predominantly Senegalese *trailleurs*) defeated an army of 12,000 Sudanese slave-raiders. In 1908, 389 French troops routed the 10,000-man army of Wadai. Perhaps the most famous of such imperialist battles was Kitchener's defeat of 40,000 Dervish troops in 1898. After 5 hours of fighting, 11,000 Dervishes lay dead, compared to only 20 Britons and 20 of their Egyptian allies.[25] The US made use of similar advantages in battles against Native Americans and, in 1898, against outmoded Spanish forces. In the naval battle of Manila, the US ships had longer-range firepower and suffered only eight casualties as they destroyed the Spanish fleet.[26] In the aftermath, other nations increased the pace of military innovation.[27] By 1914, all the major powers had repeating rifles, warships with nickel-steel armor, long-range cannons, new kinds of exploding shells, the airplane, the first crude tanks, and machine guns.

During World War I, both sides were shocked to discover that advanced technologies, when shared equally, led not to quick victory but to stalemate. This should not have been a surprise. The American Civil War had demonstrated the perils of a frontal assault against modern firepower, perhaps most famously in Pickett's ill-fated charge at Gettysburg. Nevertheless, when hostilities began in August 1914, both sides expected the conflict to last less than a year. Military experts "envisioned conflicts that turned on brief, decisive battles and heroic deeds."[28] However, in contrast to the imperialist wars, both sides had the latest technologies, which proved ideal for holding a position and slaughtering attackers. British troops sent "over the top" toward Turkish trenches were mowed down at Gallipoli as effectively as the British had killed Sudanese in 1898. In France, after more than a million casualties in the Battle of the Somme (1916), the armies were entrenched in nearly the same positions as when they had started. "The surprise of World War I," David Headrick writes, "was that the offensive had become suicidal, that vigor, élan, courage, esprit de corps, and all the other presumed virtues of European fighting men were irrelevant against a hail of bullets."[29] As Lewis Mumford reflects, "the difference between the Athenians with their swords and shields fighting on the fields of Marathon, and the soldiers who faced each other with tanks, guns, flame-throwers, poison gases, and hand-grenades on the Western Front, is the difference between the ritual of the dance and the routine of the slaughter house."[30]

The murderous efficiency of World War I showed that cavalry and foot soldiers would risk being cannon fodder in future ground engagements. In the aftermath, each military developed a distinctive technological response to this new tactical situation. In France, the superiority of defensive firepower led to the conclusion that massive fortifications along the German border would make invasion impossible. The French built the Maginot Line after adopting the doctrine of methodical battle. Victory seemed a mathematical certainty, insofar as they commanded "huge concentrations of firepower."[31] Germany made a strikingly different technological choice, favoring rapid tank maneuvers and superior communications. In 1939, the German army dashed through neutral Belgium. By going around the Maginot Line, they rendered it useless.

Another major change in warfare after 1919 was the development of air power. During the 1920s, trainees in the US Army Air Corps read Giulio Douhet's book *The Command of the Air*, which argued that future wars would be won by bombing the enemy's factories, power plants, laboratories, and supply lines. The American "shock and awe" campaign against Iraq in 2003 was nothing new. Eight decades earlier Douhet expected that precision bombing would terrorize and demoralize civilians, destroying the enemy's will to fight. To this end, American military planners put the B-17 into production in 1937.[32] They were trying to catch up with the Italians, the Germans, and the Japanese, who had used overwhelming air power against Ethiopia, Spain, and China. Picasso's painting *Guernica* protested not only the bombing of a Spanish village but also how routine such barbarism had become.

Planners originally assumed that bombers could be used to weaken the enemy, or to "soften up" well-protected positions before sending in ground troops. The British found, however, that anti-aircraft fire made precision bombing difficult. By 1941, the

Royal Air Force had shifted to night bombing to evade detection, sacrificing accuracy for crew safety. By early 1942, when the United States entered the war, British official policy had shifted from raids on strategic targets to attacks on civilians. The United States initially rejected this tactic, believing its B-17s were flying fortresses that could penetrate deep into Germany on precision bombing runs. In practice, however, 5 percent of aircraft were lost on a typical mission, and one of six never returned from some forays deep inside Germany. With such losses, most of the planes would be shot down in a month. Furthermore, these early missions did less damage to enemy factories than had been expected.[33] During the next three years, fleets of American bombers increasingly targeted civilian populations. In 1945 more than 35,000 people died in the bombing of Dresden. That same year, in the attack on Tokyo, thousands of incendiary bombs created a 2-by-12-mile firestorm that destroyed 267,000 buildings and killed everyone trapped inside.[34] Strategic air power made the most of US productive capacity, relying on expensive weapons rather than putting soldiers in harm's way. Americans had far fewer casualties than the other warring nations (292,000 in the military and very few civilians).[35]

The United States used its industrial capacity to increase air power and to reduce the need for ground troops. Indeed, a common justification for using the atom bomb against Japan remains that it rendered an invasion unnecessary and so spared the lives of a massive landing force. However, some of the scientists who helped build the bomb opposed using it on a city and proposed that it be demonstrated in an unpopulated area.[36] The Hiroshima bomb killed 107,000 people almost instantaneously, many of them by extreme heat.[37] Most were civilians, and many were children on their way to school. The hands on a young girl's wristwatch fused to the dial at the time of the blast; her blood had boiled into steam and burst out of her veins.[38]

Albert Einstein did not have such victims in mind on August 2, 1939, when he wrote to President Roosevelt that such a weapon was possible. The Manhattan Project required a new level of integration between research science and the military, as Americans developed not only the bomb but also the capacity to manage large secret projects by coordinating the efforts of thousands of specialists in all parts of the country.[39] Most had no idea of the ultimate purpose of their work. The program to build the atomic bomb illustrates how in the modern state scientific research, industrial development, and military needs dovetailed.

Since 1945, the American military increasingly has relied on technological solutions. Working together with civilian contractors and thousands of engineers and scientists, the Pentagon funds anticipatory weapons systems. They are built just in case they might be needed. Once in existence, however, they are often used. The first atom bombs were created just in case the Germans were also building them. After Germany's defeat, no one seriously argued that Japan was making atomic weapons, and the Manhattan Project might have been halted or at least subjected to democratic scrutiny and civilian oversight. Instead, the bomb remained secret until dropped on Japanese civilians. Over and over again, new weapons have been invented that were thought to be so horrible that war would be unthinkable. Weapons expected to abolish war and usher in universal peace included the submarine, the torpedo, the balloon, the machine gun, the airplane, poison gas, land mines, missiles, and laser guns.[40] Orville Wright mistakenly concluded in 1917 that "the aeroplane will help peace in more ways than one—in particular I think it will have a tendency to make war impossible."[41] He thought aerial surveillance would detect troop movements and render surprise attacks impossible. An inventor hero in a popular novel explained to his girlfriend why invincible weapons are necessary: "To have a world at peace there must be massed in the controlling nations such power of destruction as may not even be questioned. So we shall build our appliances of destruction, calling to our aid every discovery and achievement of science. When . . . war means death to all, or the vast majority of all who engage in it, there will be peace."[42]

The inventor Nicolas Tesla came to the opposite conclusion. He argued that each new weapons system, far from making war unthinkable, rather "invites new talent and skill, engages new effort, offers new incentive, and so only gives a fresh impetus to further development." After the discovery of gunpowder, "would we not have thought then that warfare was at an end, when the armor of the knight became an object of ridicule, when bodily strength and skill, meaning so much before, became of comparatively little value? Yet gunpowder did not stop warfare: quite the opposite—it acted as a most powerful incentive."[43]

Events proved Tesla correct. The atomic bombs dropped on Japan did not discredit weapons of mass destruction but stimulated their production. The Cold War drove arms development to an unprecedented level. Billions of dollars went into research and development for intercontinental ballistic missiles, new kinds of atomic weapons, longer-range bombers, spy satellites, advanced radar, computerized weapons, night vision systems, stealth bombers, precision-guided "smart weapons," superior sensors that provided real-time "battlefield awareness," and on and on.[44] It became "normal" that one out of every three American and Soviet engineers and scientists should work for the military or for a military contractor.

Viewing this arms race when it was still new, as he stepped down from the US presidency in 1961, Dwight D. Eisenhower warned against the creation of what he called a "Military Industrial Complex." He worried that the Pentagon, weapons manufacturers, and congressional representatives from districts with defense industries had become a powerful self-sustaining alliance in American politics. Eisenhower knew that new weapons have advocates and opponents whose concerns range from national security to rivalries between branches of the armed services to "pork-barrel" politics. Eisenhower worried that the proponents of new weapons systems had become such a powerful socioeconomic force that they would direct and control defense spending. He feared that the armaments industry would achieve technological momentum. A quarter-century later, George Kennan gloomily concluded that Eisenhower had been correct: "Millions of people, in addition to those other millions that are in uniform, have become accustomed to deriving their livelihood from the military-industrial complex. Thousands of firms have become dependent upon it, not to mention labor unions and communities. . . . An elaborate and most unhealthy bond has been created between those who manufacture and sell the armaments and those in Washington who buy them."[45]

Civilians, who ultimately pay for weapons systems, expect the military to protect them. But as weapons became more powerful, more civilians died in warfare, despite continual predictions of improved accuracy. In World War I, civilians accounted for 15 percent of all deaths. In World War II the rate had more than quadrupled, to 67 percent, and it has continued to rise. In more recent conflicts, notably Vietnam and Iraq, as many as 90 percent of the dead were civilians.[46] Furthermore, troops using sophisticated weapons often die from accidents and mistakes.

Eisenhower realized that inventing and deploying powerful weapons does not necessarily ensure security. Disarmament, when possible, is a much safer course of action than escalation, as the Soviet Union and the United States gradually realized during the Cold War. Their advanced weapons technologies ensured not protection but eventual doom. At the end of World War II, Lewis Mumford warned that the US government was "carrying through a series of acts which will lead eventually to the destruction of mankind."[47] Atomic weapons undermined the sense that the natural world could be taken to be the unalterable baseline of existence. As bombs proliferated, a "death world" empty of all forms of life emerged as the possible end of history. Looking back on the Cold War, George Kennan noted that Americans mistakenly had supposed "that the effectiveness of a weapon was directly proportional to its destructiveness—destructiveness not just against an enemy's armed forces but against its population and its civilian economy as well. We forgot that the aim of war is, should be, to gain one's points with the minimum, not the maximum, of general destruction. . . . We neglected to consider the strong evidence that the nuclear weapon could not be, in the long run, other than a suicidal one. . . ."[48]

At the same time, the American government sponsored nuclear power plants through the "Atoms for Peace" program. During the 1950s and the 1960s, nuclear power seemed to promise infinite and inexpensive supplies of energy. Indeed, France embraced this possibility, and made atomic plants symbols of national identity. By 2000, more than 80 percent of France's electricity came from nuclear plants.[49] In contrast, only about 20 percent of American electricity production was nuclear. This was not only because the United States had extensive fossil fuels that made nuclear plants economically uncompetitive, but also because the American public worried about potential hazards. In 1953 Edward Teller had said in a public address that "a runaway reactor can be . . . more dangerous than an atomic bomb."[50] Gradually, the public came to understand that a reactor transmuted uranium-238 into lethal plutonium. A report issued in 1957 by the Brookhaven National Laboratory estimated that a reactor accident might kill more than 3,400 people, injure 43,000, and cause as much as $7 billion in damages.[51] By the 1970s, when many nuclear plants were planned and built, few people wanted them anywhere near their homes. The accidents at Three Mile Island and Chernobyl forced the public to think harder about such dangers.

During the Cold War, the Soviet Union and the United States maintained a balance of terror, based on the shared conviction that a war could only lead to Mutual Assured Destruction (MAD). Significantly, this macabre doctrine assumed that millions of defenseless civilians on both sides would be the primary victims of nuclear war, rather than weapons systems and armies. After three decades of Cold War military research and development, weapons at times seemed to be less for national security than to satisfy the military-industrial complex. The curious result of an intensive arms race was that the military could no longer fulfill its fundamental function, to protect the population and the infrastructure.[52] MAD assumed that the army would survive, but half the civilians on both sides would perish, and radiation would make vast areas uninhabitable.

In practice, elaborate weapons systems can prove unreliable because malfunctions increase with the complexity of the system. The more parts there were in advanced jets, the more time they spent on the ground being serviced and repaired. The warning systems designed to detect a Soviet attack malfunctioned many times, including five incidents in an eight-month period in 1979–

1980. On one occasion, "a technician at NORAD—the North American Air Defense Command—accidentally placed a training tape into the main systems at NORAD's Cheyenne Mountain Complex in Colorado. That mistake made NORAD's early warning system computer think the United States was undergoing a massive Soviet missile attack. . . ."[53] Each time, central command had just minutes to decide whether or not the system was correctly advising them that World War III had begun. The American public seldom heard about these incidents, but the popular novel *Fail-Safe* described a similar self-induced nuclear catastrophe. It sold millions of copies in 1962 and became the basis for a film two years later. In *Fail-Safe,* a small malfunction—a single burnt-out condenser at Strategic Air Command headquarters—causes a computer error that sends a fleet of American bombers into Russia, destroying Moscow and forcing the American president to choose between universal annihilation and allowing the USSR to destroy New York City.[54]

But *Fail-Safe* did not predict the future. All of the presidents from Eisenhower to Reagan argued that the massive military expenditures were necessary to deter war. Neither the Americans nor the Soviets used their nuclear arsenals, and the policy of deterrence did lead to the nuclear Test Ban Treaty of 1963 and eventually to significant disarmament in the 1990s. This experience suggests that a well-armed nation may be safer than one that relies on good will and diplomacy alone. Paradoxically, some weapons are most useful if they are never used. It would seem that powerful military technology did help democratic societies to prevail against direct threats to their existence.

In contrast, limited wars suggest the perils of relying on military technology to resolve political problems. The Vietnam War proved that absolute control of the skies and massive bombing raids were not sufficient for victory. Control of the ground demanded not only firepower but also winning the "hearts and minds" of the local population. Since Vietnam, a new generation of "smart bombs" and high-tech weapons has increased the technological gap between American forces and their opponents. Victory in the 2003 Iraq War came swiftly, but the Pentagon had only sketchy plans for the aftermath, and it proved difficult to translate military control into a process of democratization. War is not a technical problem but an extreme extension of politics, which regains center stage as soon as the fighting ends.

The American military has extraordinary weapons superiority, but this cannot always be translated into political dominance. In Iraq, the United States and its coalition had complete control of air space, satellite global positioning systems that guide weapons to their targets, sophisticated tanks, a complete blanket of radar coverage, and much more, leaving no doubt about the military outcome. In Iraq, as in the imperialist battles of the late nineteenth century, there were huge discrepancies in the fatalities suffered by each side. And as in World War II, more civilians than soldiers died. There is no reason to expect this to change in the near future. A new generation of weapons promises to make war even safer for the American military. These include an unmanned six-wheel ground vehicle that can be used for reconnaissance or combat, drone airplanes that can both photograph and bomb the enemy, a machine gun that reportedly can fire a million rounds per minute, and tactical mobile robots. "There are 3-pound surveillance bots that frontline soldiers could lob through a window or around a corner to get an audio and video preview of conditions. There are robots that can negotiate harsh terrain, scurry up stairs, or rush into battle to rescue injured soldiers . . . deliver jolts of electricity, sniff for bio germs, and see through walls."[55] (But can robots be programmed to recognize and kill only enemies?) To protect soldiers further, military researchers are creating a new body armor, or exoskeleton, that not only protects but also makes it easier to carry additional equipment. It includes "a visor that expands the field of vision, as well as devices to provide information about battlefield conditions, coordinate other soldiers and decrease friendly-fire casualties."[56] As long ago as 1986, one scholar concluded that the Pentagon soon could create an "automated battlefield."[57]

Wars cannot be won by firepower, automation, and air strikes, however. As John Antal points out, "a determined enemy can always endure the fire, as the British did in the Battle of Britain, and will eventually develop asymmetrical ways to respond to precision-strike forces—as the North Koreans have with their hardened artillery sites. Precision strikes that are not backed up with a continuous battle of decisive maneuver are merely artillery raids set out to punish, not defeat, an opponent."[58] The 2003 Iraq war exemplifies superior firepower and "decisive maneuver." As subsequent events showed, however, overwhelming military power alone did not confer political legitimacy or day-to-day control.

Moreover, civilians are becoming more vulnerable in the course of everyday life. The potential dangers were suggested in 1995, when a "doomsday sect" distributed sarin gas in the Tokyo underground, and again in 2001, when anthrax spores were sent through the US postal system. The increasing sophistication of genetic engineering makes far worse things possible. In July 2002, "scientists announced the creation of a polio virus from segments of DNA ordered by mail and genetic information available on the Internet. . . . Experts say that in the near future any virus, including the most dangerous, could be produced in this way."[59] Even more alarming, it may soon be possible to produce biological weapons that target certain racial traits, for example to reduce fertility in a particular group.

The International Red Cross has become concerned about the defenselessness of civilians facing biological weapons. "Wars have been fought either with some form of safety net—however inefficient—or with some form of humanity—however selective—and with some form of humanitarian assistance to the victims—however tardy." But with the new biological weapons, "the safety nets may fall away and existing norms of law may be smashed." Indeed, unless the Red Cross and other humanitarian organizations are supplied in advance with antidotes to biological weapons, their workers will be unable to enter contaminated environments.[60]

Technological improvements have often been presented as antidotes to war. When each new form of communication was introduced, someone argued that it would increase international understanding by breaking down cultural barriers. Such claims were made for the telegraph, the telephone, radio, the motion picture, the television, and for the Internet.[61] In practice, however, each of these technologies also has become a tool of warfare. The military has used them to improve communication between the troops and their commanders, to misinform the enemy, to maintain positive public relations on the home front, and to demoralize enemy civilians.

Similarly, for two centuries new weapons have been presented as terrifying antidotes to war. Each was considered so horrible that combat would necessarily become obsolete. Today, the military can destroy all life on the planet, and the bombing of civilians has become so routine that it provokes little comment. The prestige of powerful weapons attracts more participants to the arms race. In the midst of the Cold War, Bertrand Russell anticipated these dangers. In an open letter to Nikita Khrushchev and Dwight D. Eisenhower, he warned that the escalation and spread of nuclear weapons could lead to mass extermination and even an end to human life. "There is," he wrote, "no end to this process until every sovereign state is in a position to say to the whole world: 'You must yield to my demands or die.'"[62] When Russell urged disarmament as the only rational course of action, only Britain, the United States, and the Soviet Union possessed nuclear weapons. France, Pakistan, India, Israel, North Korea, and China have now joined the "nuclear club." Other countries—notably Sweden, Germany, Australia, Canada, Brazil, and Japan—have the capability to build such bombs but have restrained themselves in hopes of preventing proliferation.

Since Russell wrote his letter, the situation has worsened because more nations have atomic weapons, because terrorists seek them, and because the bombs have become smaller. The first atomic weapons weighed several tons; by 1986 the warhead for a cruise missile weighed only 270 pounds.[63] A prominent nuclear physicist concluded "that a homemade nuclear bomb is not an impossibility, that such an undertaking need not even be particularly difficult," and that even in 1974 more than 10,000 people knew how.[64] Terrorists could drive an ordinary car containing a small "dirty bomb" to a target, and could activate it with a mobile phone.

Advanced technological systems of communication, surveillance, and defense are clearly useful for combating the threat of terrorism against civilian populations. But many terrorists are from middle-class backgrounds and have advanced education in engineering, medicine, and science. Terrorism is their immoral continuation of politics by other means. They recognize the futility of openly attacking a European or American military that has overwhelming technical superiority. The larger the technical gap between adversaries, the more tempting guerrilla warfare becomes for the weaker side. Troops with digital communications, satellite guidance systems, and miniaturized weapons push adversaries to abandon direct engagement in favor of covert attacks. The sheer superiority of Western weapons systems makes terrorism attractive to those who cannot hope to win a conventional military victory. At the same time, the complexity of Western technological society makes its infrastructure vulnerable.

To be sure, there have been some successful examples of restraint. The use of biological weapons and poison gas in World War I led the great powers to sign a treaty in 1920s that banned their military use. This ban generally has been honored, which suggests that disarmament is a surer path to safety than increasing the size of the arsenal. Likewise, after the United States and the Soviet Union had built up stockpiles of nuclear weapons during the Cold War, they agreed to destroy many of them. After four decades of nuclear stalemate, they had realized that the more military technology they had, the greater was the potential danger. Similarly, after 1946 India and Pakistan fought three wars with conventional weapons, but they learned self-restraint once they had obtained nuclear bombs.

Nevertheless, even assuming humanity avoids an atomic apocalypse, the increasing complexity of technical systems and the unforeseen accidents that may result from their interconnections point to a paradox. Even if each individual machine is becoming safer, the interconnection of many technical systems contains latent malfunctions. Likewise, weapons systems can increase national security by intimidating potential enemies, but they can also contribute to inadvertent disasters on an ever-increasing scale. As many small flaws accumulate in both civilian and military technologies, the potential for accidents increases. Paradoxically, advanced technologies can make us safer, but to do so they unavoidably expose us to new dangers.

"Faith" is a fine invention
When gentlemen can *see*—
But *Microscopes* are prudent
In an Emergency.

—Emily Dickinson[1]

It may seem obvious that technologies have expanded our knowledge of the world. The microscope revealed organisms that previously had escaped human notice. Look through a telescope and a universe appears that is invisible to the naked eye. Drop a microphone into the sea near a pod of whales and hear them sing. Insert a tiny camera inside a mother's womb and see an unborn child. Launch a satellite to Mars or the rings of Saturn and it can send back detailed photographs. Technologies enable us to see and hear things that would never be available to the unaided senses. It would seem incontrovertible that technologies on the whole increase our understanding of the universe, even if it takes time to adjust to new devices before one can interpret what they reveal.

Yet the question is not so simple. As we adapt to each new instrument and device and learn to (re)interpret the world, we may be losing touch with other modes of understanding. Furthermore, even if we can avoid such losses, an increasingly technological society may be driving us toward sensory overload. Consider multi-tasking. One often hears teachers and parents praising children for their ability to do several things at once, such as listen to music, do homework, respond to frequent telephone calls, and surf the Internet. People increasingly want continuous interactive engagement on several levels. An executive at a conference listens to speeches and takes notes on her laptop computer while also e-mailing and following the stock market. An electrician listens to the radio and answers frequent phone calls while working. Such people, and there are more of them all the time, seem almost compulsive about keeping up with information flows. A Harvard psychiatry instructor commented that the attraction is "magnetic" and compared it to a "tar baby": "The more you touch it, the more you have to do."[2] Some specialists find that many people today suffer from attention disorders and have short attention spans. "They become frustrated with long-term projects, thrive on the stress of constant fixes of information, and physically crave the bursts of stimulation from checking e-mail or voice mail or answering the phone." In some cases, multi-tasking becomes the symptom of addiction, and it may be reasonable to compare "the sensations created by constantly being wired" to "those of narcotics—a hit of pleasure, stimulation and escape."[3]

Many drivers have trouble keeping their eyes on the road as they drink coffee, eat, and talk on the phone, sometimes steering with their knees. A few drivers try to work on their personal computers or personal digital assistants.[4] Drivers have even been arrested for watching videos. At least one study has found that sober drivers using cell phones have slower reaction times and cause more accidents than drunk drivers. A psychology professor commented: "The cellphone pulls you away from the physical environment. You really do tune out the world."[5]

In the office, constantly overlapping information flows interrupt one another. Reading the conventional mail is often set aside to answer e-mails, which in turn may be interrupted by the telephone or by a breaking news story on the Internet. At least one study confirms a common-sense view of this situation: these overlaps distract people and lower their productivity. Even those who switch back and forth between just two activities, like writing and e-mailing, "may spend 50 percent more time on those tasks than if they complete one before starting the other."[6]

In small doses multi-tasking is a valuable ability, but in large doses it can lead to information overload. Ever lighter and more powerful laptop computers have wireless connections, so that e-mail, radio, television, and voice mail converge on their owners, whose mobile phones provide additional links to information flows, including music, photographs, and films. The corporations that spend billions of dollars manufacturing and marketing these machines will continue to look for ways to enhance their multi-tasking potential. Simpler machines whose "limits" might keep the user's attention more focused are disappearing from the stores. As laptops that can receive e-mail anywhere become the norm, multi-tasking turns into a "natural" part of the communications environment. With this change comes a blurring of office and home, of work time and personal time, and of news and gossip.

As more information demands attention, what will become of leisure, privacy, and personal space? Will they be submerged in endless information flows? Leisure, privacy, and personal space themselves are recent historical constructions made possible by industrialization. Leisure hardly existed as a concept for ordinary people in the nineteenth century, when farmers and factory hands worked as many as 14 hours a day. For most people, leisure only emerged as labor unions fought for shorter working hours.

Likewise, privacy was not possible for many before the second half of the nineteenth century. Before then, most houses were small, with shared public rooms and shared sleeping rooms. The idea that children could and should have individual bedrooms is at most a few hundred years old. Likewise, sitting alone and reading a book is a relatively new human activity, and it may be replaced by other forms of communication. Privacy may even go out of fashion. Sherry Turkle observes: "Today's college students are habituated to a world of online blogging, instant messaging, and Web browsing that leaves electronic traces. Yet they have had little experience with the right to privacy. . . . Our children are accustomed to electronic surveillance as part of their daily lives."[7] Perhaps this helps explain the international popularity of *Big Brother* and other "reality shows" that intrusively film the daily lives of ordinary people.

"Single-tasking" (the pursuit of one task at a time, in isolation from other people) is arguably not a natural condition of humankind but rather a possibility that emerged when industrial production generated enough surplus so that the middle class could build large houses and apartments with many individual rooms. The multi-tasking environment of constant information flows may seem un-natural, but it is no more so than a three-bedroom house for a family of four, which emerged as the norm in Western society during the last 150 years. Such houses suggest how technologies are woven into a person's conception of what is normal.

We can catch sight of the process of naturalizing technologies by looking back several generations to see how something now taken for granted was celebrated when new. In 1927 *The Nation* reflected on the spread of machines into all aspects of everyday life: "During the last quarter century machines have been carried to the farthest village in the country. Telephones, the telegraph, mechanical devices . . . yet what man thinks tenderly of his cream separator? And who looks on a telephone except with impatience and contumely?"[8] If people seemed little disposed to love most machines in 1927, the Ford automobile seemed an important exception: "Henry Ford . . . alone won the hearts of his customers."[9] In the long view, the affection for the Model T was not a fixed emotion, however. Rather, every generation of Americans has lovingly embraced a machine whose advantages their descendants took for granted while complaining about its service. The railroad seemed magnificent in 1840, but appeared to be a grasping monopoly by 1890. Airplanes miraculously conquered gravity in 1910, and pilots were demigods. But these emotions cooled as flight became routine, and passengers focused on the tedium of checking in, on leg room, on airline food, and on lost luggage. Indeed, the very technologies that *The Nation* listed went through the same process of banalization. The first people to use the telegraph, and later the first to use telephone, celebrated them as marvels that overcame time and space. Cream separators also seemed wonderful to those who first used them. The emotional response to technologies changes over time. As people weave particular machines into the texture of everyday life, they cease to be the center of conscious attention. Railroads that run on time and perfectly functioning electrical grids don't draw comments; they have become "natural."

Such "naturalization" has been going on for thousands of years. In "The Poet," Emerson chided his contemporaries for thinking that industrialization was not in harmony with nature: "Readers of poetry see the factory-village and the railway, and fancy that the poetry of the landscape is broken up by these; for these works of art are not yet consecrated by their reading; but the poet sees them fall within the great Order not less than the beehive or the spider's geometrical web. Nature adopts them very fast into her vital circles, and the gliding train of cars she loves like her own."[10] Others also saw technological achievements not as violations of nature, but rather as extensions and completions of it. To the nineteenth-century American, "man's conquest of the mountains was not a violation of Nature but an embrace."[11] The process of naturalizing technology begins with the assertion of harmony between a machine and the natural world, and ends with its unconscious acceptance by later generations, since the machine was already there when one was born.

José Ortega y Gasset concluded that "technology is man's reaction upon nature or circumstance" and that it "leads to the construction of a new nature, a supernature interposed between man and original nature."[12] Before long, this "supernature" begins to seem natural. "Since present-day man, as soon as he opens his eyes to life, finds himself surrounded by a superabundance of technical objects and procedures forming an artificial environment of such compactness that primordial nature is hidden behind it, he will tend to believe that all these things are there in the same way as nature itself is there without further effort on his part: that aspirin and automobiles grow on trees like apples."[13]

Musical recordings exemplify this point. Today's child listens to a compact disk and hears a rich tapestry of sounds that is not merely amplified but transformed through digital manipulation, producing music that cannot be heard in any concert hall. The sounds on the child's CD are technological in other senses as well. For thousands of years, all music except human singing had been produced by complex man-made mechanisms, notably flutes, horns, trumpets, drums, violins, guitars, and pianos. Music could not be reproduced until 1877, when Edison invented the phonograph. Early witnesses were amazed by Edison's recordings, but later generations found them inferior to what they copied. For about 100 years, the goal of recording engineers was to make a perfect copy that was indistinguishable from the sound of a live performance.[14] In the last generation, however, the sound of recordings has increasingly been enhanced. Instead of a single microphone carefully placed to capture a representative listener's experience from the best seat in the house, technicians placed many microphones in the hall and then modulated and mixed the sounds in order to bring out certain instruments, textures, and effects. The music they produced no longer represented what anyone could hear from any seat in the concert hall.

The sound the child enjoys arrives through the mediation of a laser that reads the digital code on the CD and transmits it to the amplifier, which in turn sends a signal to the speakers or headphones. Although every CD is identical, the sound produced varies somewhat, depending upon the quality of the machine used and the settings selected. In the pre-digital days, sound engineers and consumers constructed a shared definition of "high fidelity." But in the case of a CD heard in 2005, fidelity to what? There is no original sound being copied. Rather, there are elements being assembled and manipulated to create a pleasing effect that may resemble what one might hear if allowed to roam unobtrusively around a concert hall, so as to be close to the flutes, the cellos, or the oboes during their solos but at the center of the hall for the crescendi and the grand finale. The manipulation of sound goes far beyond the placement of microphones, however, as a glance at the hundreds of buttons and knobs on a studio's sound table suggests.[15] Equally skilled manipulation occurs in film, in television, and in computer animation, as indeed it did in old-fashioned analog photography. Who has not arrived at a tourist hotel and realized how cleverly a photographer has made rooms appear more spacious and comfortable than they are?

What does such a world of perception imply? How do such manipulated sounds and images affect us? Are we lost in a hall of distorting mirrors and loudspeakers, or have we moved into a new world of post-natural perceptions? Martin Heidegger complained: "But we do not yet hear, we whose hearing and seeing are perishing through radio and film under the rule of technology."[16] Heidegger distrusted manipulated sounds and images, standardized and stockpiled for sale. Like him, many still prefer direct experience of a live musical performance without the many levels of technological mediation.

But the concert hall itself is a technological artefact, invented and perfected over centuries to house orchestras and their audiences. Architects learned to design these structures to accommodate the eye and (even more important) the ear. Audiences demanded halls with clear, lively acoustics that were not muddied by echoes and yet had sufficient reverberation to carry sounds to the furthest corner. The most successful halls, such as Carnegie Hall in New York, are not merely beautiful to look at and acoustically perfect; they also have good lighting and unobtrusive heating and ventilation, so that the audience experiences no environmental distractions. Such halls remain surprisingly rare, however, and even prestigious projects such as Lincoln Center in New York often have inferior acoustics when completed. James Marsden Fitch concluded that "the technical means at the disposal of the architect are incomparably higher than ever before, [yet] he is producing new theaters whose overall performance is less satisfactory than many built centuries ago."[17] In part, this is because many people shape the final result, including specialists in acoustics, lighting, air conditioning, and building. This specialization can undermine a unified approach to the experience of architecture, often because the visual becomes more important than the aural and the tactile. Philharmonic Hall at Lincoln Center looked spectacular, but it proved so unsatisfactory to musicians and audiences alike that it required alterations. Both the instruments used to produce the music and the space in which it is heard are technological constructions. It is naive to condemn manipulation of the sound on CDs, since live classical music also requires a sophisticated technological support system.

In the late 1980s, when Tracy Chapman became popular, it was in part because she sang alone and played an acoustic guitar with only simple amplification. Her "unplugged" sound seemed fresh and authentic, in contrast with the more obviously contrived work of other musicians. However, to reach audiences, her songs also required the elaborate technologies of recording and distribution used in all other popular music. From a Heideggerian perspective, Chapman's "authenticity" was only a disguised artificiality.

The problem of authenticity also haunted early users of the telephone, who worried about the identity of the disembodied other at the end of the line. Inexperienced users were at times hoaxed or cheated, as confidence men used the telephone to extract from the unwary information, money, or agreements to meet, much as has occurred more recently over the Internet.[18] Aside from these obvious forms of misrepresentation, the telephone and other forms of disembodied communication present a more pervasive problem of uncertainty. In "The Neighbour" and The Castle, Franz Kafka explored the telephone's potential for "schizophrenia, paranoia, dissimulation, and eavesdropping."[19] Severing the human voice from physical presence creates unanticipated possibilities.

Such potential uncertainties are often invisible. In everyday life, technologies mediate almost all experience from the moment one awakens until going to sleep at night. Much of what one sees is subtly shaped by the spectra of light thrown by different types of bulbs and fluorescent tubes. The air itself is heated, cooled, or dehumidified according to the needs of the location and the season. What one hears is muffled, amplified, or otherwise mediated by man-made materials, and a good deal of this sound is transmitted by radio, stereo, television, computer, or telephone. The shape, texture, and taste of the orange juice, eggs, coffee, and English muffin one eats for breakfast have been modified by a myriad practices, including the breeding and feeding of animals, the use of food additives and preservatives, and the transformation of raw foodstuffs into products at processing plants. When leaving in the morning, few people directly experience much of the weather; they see it through the windows of cars, buses, and trains on the way to school or work, where "reality" is increasingly defined by telephones and computer screens. As the torrent of signs and representations sweeps people along, does the unitary sense of self break down? Todd Gitlin has argued that in a heavily mediated world "each person will be multiple. Each will feel disordered and restless. Each will be comfortable relating to, feeling with, trying out the most accessible repertory of stories and sounds, cutting and pasting, surfing and clipping. Each will sprout multiple auxiliary relationships to figures who never breathed."[20]

At what moment in a typical day does one have an unfiltered experience in which technology plays little or no part? A walk in a park during lunch hour? But a park is a technological artefact, especially if it has a lawn. Lawns are tended by machines. The clipped, even surface of the ideal lawn is the result of pesticides, fertilizers, weed killers, and regular mowing. Such lawns, largely invented in the nineteenth century, have been widespread in Western culture for a relatively short time.[21] Likewise, it is highly unlikely that all the plants in a park are native to the region, or that they will thrive there without pruning, fertilizer, and occasional spraying.

Perhaps a natural, unmediated experience can be found in a forest or a national park? At the Grand Canyon, the waters of the Colorado, once warm and red, are now cold and green, because the rocks and sediment that the river once carried downstream have settled behind a series of man-made dams. Much water is siphoned off for irrigation, the remaining current in the main channel is too weak to wash away sediment deposited by tributaries, and "deposits up to 2.6 meters thick have accumulated within the upper Grand Canyon."[22] The shoreline, no longer subject to spring floods, has a more luxuriant plant life than before, including millions of tamarask plants, which are not native to North America, and these in turn provide shelter for many birds. Seen from the rim of Grand Canyon, these plants do not seem like interlopers, but rather suggest an edenic oasis.

Tourists' expectations increasingly seem to emerge from experiences of film and television, as is suggested by a 1995 Gary Larson "Far Side" cartoon. A couple stands at a "canyon lookout." The husband says: "I dunno. We're just so far up, I think this'd be better on the tube." In fact, a 1992 Sony commercial showed a large television set perched near the Grand Canyon rim. A small boy ran to it, more intrigued with the canyon's image on TV than with the actual landscape.

Few of the tourists who visit the Grand Canyon actually descend all the way to the bottom, but every year more than a million go to the IMAX theater just outside the park's entrance. It has quadraphonic sound and a screen the size of a two-story house.

The 34-minute film represents a descent to the river, a flight over the canyon, and a river raft trip; the raft ride is advertised as more dizzying and terrifying than the real thing. IMAX theaters also have been erected at Niagara Falls and at Yosemite. The commercial success of such facilities right at the entrances to the sites they represent suggests that tourists want an intensity and an enhanced "authenticity" much like that of a television program or a music recording. In each case, the consumer perceives the object from multiple angles, hearing and seeing in ways that would be difficult or impossible in the "real world." These tourists seem to prefer mediated experience to direct use of their senses. As the examples of music, tourism, and a walk in the park suggest, "the devices, techniques, and systems we adopt shed their tool-like qualities to become part of our very humanity."[23]

Are modern men and women so thoroughly adapted to their technological society that they cannot step back from it? Is our acculturation irreversible and inescapable? Many flee the sensory overload of urban life. Some hike into wilderness areas, escaping from all forms of technological communication and immersing themselves in unspoiled surroundings. Some have even attempted to escape from the present by living in and using the artifacts of a physical space reconstructed or preserved from another moment in history. By living in a Stone Age village or a Victorian house, they separate themselves from the world of perceptions they were born into and challenge unexamined assumptions about the 'natural' textures, tastes, sounds, and appearances of daily life.

By the early years of the twentieth century, city dwellers realized that a fundamental rupture had separated them from nature, putting in its place an artificial environment of skyscrapers and of subways that ran day and night on electricity. In 1909, E. M.

Forster imagined the ultimate result of substitutions for direct experience in "The Machine Stops." In this short story, all people live in identical single rooms under the ground. These rooms cater to every desire. At the touch of a button, water, synthetic food, music, or communication is effortlessly and instantaneously available.[24] In this world, the exchange of ideas is far more important than direct experience, and its denizens live alone. They dislike seeing one another in person, and they avoid physical intimacy. Instead, they continually exchange messages over the wires or speak to one another on screens. The sounds and the images are slightly imperfect, but that has long since ceased to bother anyone. "The Machine," a vast interlinked system powered by electricity, regulates all exchanges of information and filters direct experience. It also restricts direct contact with the surface, where few have ever been and where almost no one desires to go. Indeed, few people want to travel, since everywhere is like anywhere else. Not only is all experience mediated through the machine; the world inside it has become completely naturalized. The inhabitants of this future society disparage immediate experience or physical sensations; they prefer replications of events and refinement of ideas. They value secondhand knowledge more highly than firsthand, and the credibility of an idea increases the more it is refined through reinterpretations. Over several generations, lacking interest in the physical world, the citizens gradually have forgotten how their vast infrastructure works. When it begins to break down, they are unable to make repairs, and they begin to perish en masse. Only a hitherto unsuspected remnant of humanity survives. They have maintained a far less technological existence on the surface.

Since Forster wrote "The Machine Stops," telephones, radio, television, videos, and the Internet have become almost universal in the Western world. In 1968 one of the chief administrators of the US Defense Department's Advanced Research Projects Agency (ARPA) declared: "In a few years, men will be able to communicate more effectively through a machine than face to face."[25] In fact, a few cyber-hermits already exist. Since 1976, one man has lived alone in the Oregon woods, making a living by writing programs for Apple Computer and other companies.[26] With the spread of e-mail, electronic hermits may have increased, though taking a census is difficult.

"The Machine Stops" may now seem a fable about a world in which all people have become isolated nerds addicted to the Internet, but its fundamental theme was not new even in 1909. In the nineteenth century, Thomas Carlyle and others worried that machines would rob people of an organic relationship with nature. But concern intensified in the twentieth century. What are the psychological effects of living within a technological society that wraps humanity in a cocoon of machines and conveniences? In Max Frisch's 1957 novel *Homo Faber,* a diary entry by the engineer narrator records a conversation with his ex-wife: "Discussion with Hanna—about technology (according to Hanna) as the knack of so arranging the world that we don't have to experience it. The technologist's mania for putting the Creation to a use, because he can't tolerate it as a partner, can't do anything with it; technology as the knack of eliminating the world as resistance, for example, of diluting it by speed, so that we don't have to experience it. (I don't know what Hanna means by this.) The technologist's worldlessness."[27] The passage communicates both Hanna's critique and her former husband's failure to understand her. She is an expert on ancient Greek society, and the contrast between classical and modern civilization informs her observations. He has worked in Latin America and Mexico, installing electric power stations, which help to "eliminate the world as resistance." Electric light eliminates the night, air conditioning eliminates climate, and electric devices replace physical labor (such as pumping water or grinding wheat into flour). As Hanna complains, the result of these and many other technologies is that no one directly experiences the world.

Like Hanna, philosophers have tended to emphasize not particular machines but rather whole systems that foster technocratic sensibilities. Martin Heidegger argued that in the modern world technology provides a pre-theoretical "horizoning of experience." As technological rationality becomes dominant, people begin to perceive all of nature as a "standing reserve" of raw materials awaiting use. The transformation of this standing reserve for our comforts becomes "natural." A child born since 1950 finds it "natural" to use electric lights, to watch television, to ride in an automobile, and to use satellite-based communications. That child's grandparents regarded such things as remarkable innovations that had disrupted the "normal." In contrast, the child unquestioningly accepts these technical mediations of experience as the pre-theoretical "horizon" of perception.

Some recent philosophers of technology argue that "as devices replaced practices, disenchantment set in."[28] As the technological domain encroaches on or mediates all experience, it overtakes and delegitimizes both traditional society and older perceptions of the world. In place of the familiar cycles of everyday life in a more direct relationship with nature, "technological character is concentrated in its liberating powers to be anything, that is, to be new, to never repeat itself."[29] But this is only an apparent liberation that comes at a cost. The penetration of technology into all aspects of being means that "our new character is grounded in human-technology symbiosis," and that "prior to reflection, technology transforms character."[30] The transformation imposes itself on each child, redefining every generation's social construction of "normal experience."

Building on Heidegger, Albert Borgman's *Technology and the Character of Contemporary Life* posits a division between engagement with things, such as a stove, and using devices, such as a furnace. Only things engage people in a world of physical action. To use a stove, wood has to be sawed, carried, and burned, in rhythms shaped by the weather and by food preparation. In contrast, a furnace regulated by a thermostat makes few demands on skill, strength, or attention, and it becomes an unconscious element in daily life. Such devices create a sharp dichotomy between the surface and the increasingly unfamiliar machinery beneath it. Computers offer a suggestive example of how the surface emerges. The first generation of personal computer owners sometimes built their machines and usually understood how they worked. With each succeeding generation, however, computer owners are less likely to understand the insides of their machines, which have become as opaque as the automobile, the automatic furnace, the birth-control pill, the pacemaker, and thousands of other technological devices. As they become "normal," technical systems disappear into "black boxes" whose inner workings we neither think about nor understand. People accept automatic technical control, and the opacity of the machine becomes "natural." "Under the sway of technology," one post-Heideggerian philosopher concluded, "we understand the challenge of our age as one of getting everything under control as resources. Such an understanding and approach impoverishes what it touches: the farm field, the river, the forest, and even the planet itself." Technological people, in this perspective, "are always oriented in the world as challenged forth by technology."[31]

If a web of controlling technical processes and devices defines "normality," and if this web becomes more complex with each generation, can mankind depart too far from nature? Are new forms of constructed "normality" psychologically healthy? How malleable is "human nature"? Postmodernists such as David Harvey emphasize how the acceleration of transportation and communication has compressed space, sped up time, disconnected voice from presence, subverted social boundaries, intensified the circulation of information, and created a blizzard of representations. By the 1990s it had begun to seem self-evident that only a new, postmodern self could accommodate this new infrastructure. Some have speculated that "the reconfiguration of the relations between the senses, especially of hearing, seeing and touch, promised by new communicative and representational technologies may allow for a transformation of the relations between feeling, thinking, and understanding. . . . The sheer overload of sensory stimulus required to absorb by eye and ear results in a switching or referral of senses. These contemporary synesthesias make it appropriate for us to think in terms of visual cacophony and white noise. . . . We must also expect a redistribution of the values previously sedimented in the senses of hearing, vision, taste, touch, and smell."[32] If this analysis is correct, then the experience of being alive, to the extent that this means experiences we have through the senses, has already changed fundamentally, cutting us off from the ways that people saw the world for most of human history.

If a highly technological sensory world is fostering new kinds of synesthesia, some embrace this transformation. They are fascinated by the possibility that people using computers could induce fundamental changes in themselves and invent new personality structures through immersion in chat rooms, MUDs (multiple-user dialogues), and other Internet discussion forums. At the very least, individuals could explore alternative sides of themselves on the Net. An old person could pretend to be young, or a woman might assume a male name and male verbal behavior. People who assume such disguises may resemble patients with multiple personality disorders (MPD), and yet they usually can distinguish between their various online personas and a reasonably stable offline personality.

The question becomes "What sort of personality is normal in the online situation?" During the nineteenth century and well into the twentieth, the ideal personality was assumed to be "well integrated." A sane person had a strong sense of self and a stable identity. The classical psychology of Freud or Jung emphasized the desirability (indeed, the normality) of integration, not of pretending to be someone else, or participating in a consensual group hallucination. The Internet has challenged this holistic view of the self, a theme that Allucquère Rosanne Stone explored in *The War of Desire and Technology*.[33] From her perspective, the ambiguity of online gender and the development of multiple personalities are not signs of mental illness but a liberation from patriarchal structures of authority. Stone's arguments develop out of post-Freudian psychology—particularly that of Jacques Lacan, Gilles Deleuze, and Felix Guattari, who decenter the idea of the bourgeois self.[34] Stone argues that people in chat rooms who are exploring what it means to be of the other sex develop a state of mind that resembles MPD. She thinks they are by no means ill. Rather, she suggests, the Internet is allowing them to manifest a healthy plurality that most people have been socialized to repress.

Sherry Turkle takes a somewhat different view of the same phenomena. After interviewing people who spend hours online in chat rooms and MUDs, she concluded that, although they may appear to have some of the characteristics of people with MPD, there is a crucial difference. One male subject, known as "Case," likes gender swapping and often has a "Katherine Hepburn" persona when MUDing. Unlike someone with MPD, however, Case is aware of his different personas; the "inner actors are not split off from each other or his sense of 'himself.'"[35] In this view, chat rooms and role-playing games are not psychologically harmful; rather, they give people "a time out or safe space for the personal experimentation that is so crucial for adolescent development." However, creating screen identities does not prepare one to share and develop feelings in direct contact with other people. It can provide an "illusion of companionship without the demands of friendship."[36]

The possible psychological effects of immersion in the Internet must be understood against the background of the scientific debate on MPD (or Dissociative Identity Disorder, as it has also been called since 1994). Skeptics such as Ian Hacking and Elaine Showalter question the reality of MPD.[37] They note the sudden increase in its incidence and cite the fact that its seems to spread wherever therapists who have been trained to look for it are hired. One influential proponent of MPD believes that more than 2 million Americans "fit the criteria for being a multiple personality."[38] Critics charge that MPD is a social construction, developed by patients in response to the questions and the cues of the therapist. They are convinced that many of the supposed memories of childhood sexual abuse (which often is supposed to trigger the splitting up of the personality) never in fact took place, but were invented as part of the discourse of therapeutic treatment.

If these critics are correct, the conclusion seems to be that the social construction of multiple selves easily can spread through the use of the Internet. The psychiatrist Harold Merskey concludes that MPD is produced by "suggestion, social encouragement, preparation by expectation, and the reward of attention."[39] Chat rooms deliver all of these. The anonymity of the chat room and the fact that its communications are textual constructions makes it an ideal setting for MPD, especially if one can develop this condition without having suffered childhood sexual abuse and without having had several personalities ever since. The director of Johns Hopkins University's department of psychiatry declared that MPD is "promoted by suggestion and maintained by clinical attention, social consequences, and group loyalties." What better place to develop and maintain MPD than the Internet, where no one will likely be in a position to discredit new personalities, and where being consistent in a new role creates social acceptance? If MPD is more invention than mental disorder, then online it can flourish and assume ever-new guises without the assistance of a physician.

If human psychology might change through intensive use of the computer, equally fascinating is the possibility of artificial intelligence (AI). For more than 50 years, philosophers and computer specialists have argued about what would constitute AI. Clearly it requires more than the ability to calculate or to answer specific questions. One common definition (the "Turing test") is that AI requires that an ordinary person cannot tell that he or she is conversing with a machine. Others argue that to be human is to be embodied and gendered, and that therefore only intelligent machines inside recognizable bodies can achieve AI. In the 1960s, for example, Philip K. Dick, in *Do Androids Dream of Electric Sheep?* (the novel that became the basis for the film *Blade Runner*), imagined androids that were difficult to differentiate from people.[40] Dick's androids are still machines, but they have such human-like thought processes and emotions that often only a specialist can discern what they are. The notion that artificial life could have emotions is not new. The monster in Mary Shelley's *Frankenstein* experienced fear, anger, and sentimentality, and the Tin Man in L. Frank Baum's *Wizard of Oz* yearned for a heart. Likewise, in science fiction films, Vivian Sobchak found that in recent films "alien others have become less other—be they extraterrestrial teddy bears, star men, brothers from another planet, robots, androids, or replicants."[41]

In contrast to androids, cyborgs fuse the body with the machine. In one sense, this idea is as old as artificial limbs, false teeth, or eyeglasses. People have long manufactured replacement parts to make up for their defects. In such cases, a person is almost "as good as new." But the cyborg improves upon the normal human. The goal is no longer merely to be as good as a healthy person; now it is to surpass normality and to become stronger, faster, or more intelligent. In the stories and novels of William Gibson, machines are grafted to the body to create a male who has faster reaction time or increased memory or a female who has retractable razor-sharp nails and superb artificial vision.[42]

And the cyborg of fiction has a different psychology. As Donna Haraway emphasizes, "the cyborg is a creation in a post-gender world; it has no truck with bisexuality, pre-Oedipal symbiosis, unalienated labor, or the seductions of organic wholeness."[43] Its purposes cannot be understood using Freudian psychology, and it cannot yearn to return to Eden. Rather, the cyborg wants to escape the "meat body" into the matrix of electronic communications, literally to become disembodied. Gibson imagined a future world in which it has become routine to enter an electronic matrix whose roots are in "primitive arcade games" and in "early graphic programs and military experimentation with cranial jacks." Gibson's characters can venture out of their bodies and physical surroundings into "Cyberspace," which he famously defined as "a consensual hallucination experienced daily by billions of legitimate operators, in every nation, by children being taught mathematical concepts. . . . A graphic representation of data abstracted from the banks of every computer in the human system. Unthinkable complexity. Lines of light ranged in the non-space of the mind, clusters and constellations of data. Like city lights, receding. . . ."[44] In this vision, the human nervous system is hard-wired to the computer web, and Gibson's characters can voyage through the net, leaving their bodies behind.

Several films explore a related idea: the neurological replication of personal experience. In Kathryn Bigelow's *Strange Days* (1995), a new (and illegal) technology makes it possible to record the sights, sounds, sensations, and reactions that a person has during any segment of his or her life. Anyone else can play back the recordings of these experiences, much as one can play a CD-ROM. Because the recording and playback mechanism is the human body, the experiences are so powerful and vivid that they seem entirely real. People easily become "junkies" who prefer to "experience" other people's intense moments rather than go through their own daily routines. Related notions are taken up in David Cronenberg's *eXistenZ* (1999), which focuses on computer games that plug into the spinal cord and provide simulations that are indistinguishable from "real" experience.

In such visions, the technological world has become an all-encompassing blanket of sounds, texts, and representations that define reality without even entering through the old-fashioned portals of the five senses. Gone are Heidegger's anxieties about authentic experience. In Gibson's works, a fabricated consciousnesses can be invented and exist entirely on the net, become a multimedia star, and marry a human being.[45] In the cyborg psychic economy, all experiences are constructed. Life inside cyberspace not only has the most value; it is also the realm of financial and political power that defines and mediates human circumstances. Gibson's heroes accept an entirely technological world, but they resist powerful institutions that control most of cyberspace. They struggle to be autonomous virtual consumers as they surf a contested and shifting perceptual terrain that they try to shape to their will. Nature seems to have been surpassed, but it remains elusively at the edge of consciousness—until the power fails.

Gibson is not describing an inevitable future, for we are not entirely trapped inside a prison of technologically shaped perceptions. Some experiences are less mediated by machines and technical systems than others, and we can tell the difference. We can distinguish a live performance from one on a CD, and we can distinguish an invented online personality from a personality developed and shared with relatives and close friends. One need not be a trained philosopher to grasp the difference between building a wood fire and relying on automatic heating. The widespread yearning for contact with remote and uninhabited places or with wilderness strongly suggests that millions of people do not snuggle comfortably inside the cocoon of a technological world. Rather, they oscillate between embracing its conveniences, even wallowing in its pleasures and fleeing its sensory overload or rejecting its inauthenticity. Cyberpunk film and fiction express only one side of this equation.

Not Just One Future

Several of the most widely read novelists of the first half of the twentieth century evoked the terror of living in a society where technologies became the basis of massive state control. These included the dystopias of H. G. Wells (*When the Sleeper Wakes*), Aldous Huxley (*Brave New World*), and George Orwell (*1984*).[1] Reading their work is another reminder that technologies matter, whether one is looking at human evolution, at cultural diversity, at employment, at government, at consumption, at the environment, at safety, at the military, or at the overarching social construction of reality.

Reading such works also raises the question "Where does the human infatuation with technology lead?" Even the simplest tool implies a narrative that points toward the future. The chimpanzee with his peeled stick looking for termites has begun to think of a sequence in time, as hunger prompts him to make a tool with which to find, catch, and eat termites. The prehistoric toolmaker patiently chipping out arrows or axeheads conceived of a generalized future need and prepared against the day when tools should be at hand. Complex technologies demanded more resources and lengthened the gap between making and possible use. In some cases, the future benefit is obvious—for example, digging a well, throwing a bridge across a river, or building an aqueduct into a city. Yet the largest projects often lacked such practical justifications. The builders of Stonehenge, the Pyramids, or the medieval cathedrals could dedicate a lifetime to a great technological project for its religious meanings. To mobilize such energies required a sense of order that placed such buildings at the center of a whole scheme of life. Technology is not something that comes from outside us; it is not new; it is a fundamental human expression. It cannot easily be separated from social evolution, for the use of tools stretches back millennia, long before the invention of writing. It is hard to imagine a culture that is pre-technological or a future that is post-technological.

From the vantage point of the present, it may seem that technologies are deterministic. But this view is incorrect, no matter how plausible it may seem. Cultures select and shape technologies, not the other way around, and some societies have rejected or ignored even the gun or the wheel. For millennia, technology has been an essential part of the framework for imagining and moving into the future, but the specific technologies chosen have varied. As the variety of human cultures attests, there have always been multiple possibilities, and there seems no reason to accept a single vision of the future. Those who think machines are taking humanity somewhere in particular probably are wrong. No determinism made the automobile an inevitable choice instead of mass transit. Nothing inherent in technologies dictates that people should live in apartment buildings, semi-detached dwellings, or single-family houses. Nothing makes e-mail an inevitable cultural choice instead of the telephone, the old-fashioned post, or the Amish preference for face-to-face communication. Rather, each group of people selects a repertoire of techniques and devices to construct its world. A more useful concept than determinism is technological momentum, which acknowledges that once a system such as a railroad or an electrical grid has been designed to certain specifications and put in place it has a rigidity and direction that can seem deterministic to those who use them.

Neither the technologies of the future nor their social uses are predictable. Although there is a great deal of speculation in this area (including stock market speculation), even experts with impeccable credentials are often wrong. Predictions about such fundamentally new inventions as the telegraph, the electric light, the telephone, and the personal computer were little more accurate than flipping a coin. More surprising, even modifications of existing inventions are hard to predict. Many people, notably the inventors and investors who have a stake in the outcome, propose dramatic future scenarios in which their particular device will become indispensable for the average person, so that no home should be without one. This proved true for the light bulb and the radio, but not for the picture phone or a myriad other devices.

Historians, who know more of the story and who do not have a financial stake in the outcome, are more likely to get it right. Traditionally, they are divided into Internalists and Contextualists.[2] Internalists focus on how machines come to be. Fifty years ago they celebrated individual inventors, gazing over their shoulders in admiration. However, further research has shown that inventors draw on networks of people. Far from acting alone, they coordinate and synthesize. Each new machine emerges from and is shaped by the time, the community, and the place of its making. Internalists do not treat a technology as a "black box" whose inner workings can be taken on faith. They delight in opening up the machine to scrutiny, studying precisely how it worked and what new problems it presented. New machines are not simply products of laboratories, however. They emerge within shaping political and social contexts, and they can be used for many different ends. Contextualists focus on how new machines are incorporated into society. They reject the idea that the public desires unknown machines ahead of time and somehow knows what to do with them when they are invented. Contextualists seek to understand the perspectives of people in the past, most of whom were not engineers or inventors. Any new device had to prove itself in everyday life, and in many cases (notably those of the telegraph and the phonograph) an invention first seemed a novelty with few obvious practical uses. It took railroad managers a generation to see the benefits of the telegraph lines that ran along their own right-of-way. It required a generation for the public to find that it wanted the phonograph not primarily to dictate letters or preserve voices for posterity but to play music. Because historians of technology are familiar with many similar cases, almost none agree with externalists (such as Alvin Toffler) who treat machines as powerful and well-functioning "black boxes" that irresistibly transform the world around them. Such approaches generally are misleading.

Since technologies are not deterministic, it follows that people can use them for many ends. For much of the nineteenth and the twentieth century, sociologists and historians assumed that the machine age could only lead to a crushing homogeneity. But in practice, people have often used technologies to create differences. Consumers generally prefer variety. Even a manufacturer bent on absolute uniformity, such as Henry Ford, eventually had to give in to the public's demand for a range of models and options. Likewise, homeowners proved adept at transforming Levittown's rows of identical, mass-produced homes into variegated neighborhoods. Difference triumphed over uniformity.

By the end of the twentieth century, it was clear that highly technological societies prefer to maximize differentiation. Racial, regional, and ethnic communities developed separate identities by inventing and disseminating new traditions. In contact zones between highly technological societies and developing nations, a creolization process took place as many peoples selected and rearranged elements of Western culture and absorbed them into their own traditions. Technologies were not simply being used to eradicate cultural differences and create a single, global culture. In many cases they enhanced the possibilities for the survival and growth of marginal communities or minority cultures. This pattern emerges even in the urban geography of the United States. The spatial formation of Los Angeles, as analyzed by Mike Davis in *City of Quartz,* is highly suggestive.[3] Wealthy gated communities define a complex of global values in their cable network televisions, broadband Internet connections, and new automobiles. Beyond their gates are ethnic communities and poor neighborhoods, which are hardly the passive recipients of whatever trickles down from the globalized world of the movie studios and the international corporations. Rather, these local groupings are highly inventive, creating new clothing styles, musical forms, and other cultural products that often are reappropriated into "global" culture. Using mobile phones, they cruise and control much of the city. This dynamic between what John Fiske calls an imperializing power and the localizing powers of the margins energizes much of popular culture, moving it away from homogeneity toward differentiation.[4]

The expansion of multicultural consumption across the globe puts pressure on natural resources. Although extraction and production techniques have become more efficient, continual abundance is not ensured. Technological improvement does not automatically lead to long-term economic growth. People can choose wasteful methods that bring high agricultural yields but hasten soil erosion. They can build power plants and factories that produce inexpensive goods but pollute the air, leading to acid rain and deforestation. They can choose to recycle metals or not, to use powerful pesticides such as DDT or not, to allow genetically modified plants into their environment or not. Overall, each culture chooses how large a "footprint" it will leave on the land and whether it will live within limits set by its environment or treat nature as a stock of raw materials. The philosopher Holmes Rolston speculated that "in the increasingly technologically sophisticated world of the future" nature will no longer constitute the framework of life. Instead, "nature will become not so much redundant as increasingly plastic."[5] Yet treating nature in this way may have limits.

The OECD nations use many more resources than the rest of the world. In 1990, a North American family of four consumed as much power as an African village of 107. A North American used twice as much energy as a European and ten times as much as a Latin American.[6] For optimists who believe in more growth, such disparities point to new markets. But if, as seems likely, there are environmental limits to growth, then the Western world, particularly the United States, must reduce per-capita resource consumption. To judge by the political parties' programs, however, most people refuse to believe that there are imminent limits to growth and do not fear global warming. Voters support leaders who stimulate the economy, and they reject restrictions on technological innovation. In the 2000 presidential campaign, George W. Bush denied the existence of global warming. As president, he pushed to permit greater use of methyl bromide, a pesticide that depletes the ozone layer and that was in the process of being phased out.[7] Bush asked for oil drilling in the Arctic National Wildlife Refuge. He proposed weakened air quality standards that would triple toxic mercury emissions and would increase sulfur emissions and the smog-forming nitrogen oxides in the air. Bush understood that technological systems are social and political constructions; he did not believe that the environmental system had reached its carrying capacity.

Some technological liberals acknowledge the dangers of pollution, energy shortages, and resource depletion but argue that if we have invented a problem we can invent its solution. If twentieth-century automobiles were inefficient and polluting, new hybrid cars can be three times as efficient. They can use electric motors in the city and gasoline or natural gas engines on the open highway. Universally adopted, they would reduce the demand for oil by more than a half. In practice, however, average fuel efficiency in American automobiles has worsened since 1988, when it was 22.1 mpg, falling to 20.4 mpg in 2002.[8] Since some cars now on the market get more than 60 mpg, this is clearly a cultural choice. Fortunately, energy savings have been more effective in other areas. The best new washing machines use half the water and half the electricity of models from the 1980s. Likewise, better house-building techniques, including more wall insulation, thermal windows, common walls with neighboring buildings, and passive solar heating, can halve a home's energy needs. When such technological choices are combined with intensive recycling, a society might achieve the "sustainable development" called for in the "Brundtland Report" (*Our Common Future* 1987). Even if the affluent become more efficient, however, the poor will not be interested in preserving the environment if they remain trapped in poverty. Sustainability requires both lower consumption in wealthy nations and improved living standards in less developed countries.[9]

In theory, "soft" or "appropriate" technologies could be used to scale back resource demands and energy consumption. But in practice, "appropriate" technology may fall short. Solar panels can convert sunlight into electricity, but for 30 years their energy has been more expensive than fossil fuels. Even if an OECD country wanted to convert to renewable energy, solar and wind technologies could supply considerably less than half of the current they demand. Whatever the technology, social factors may prevent its adoption. In India, biogas generators that convert animal dung and organic waste into methane proved so expensive that only middle-class farmers could afford them. In the United States, the ideology of individualism thwarts some collective solutions, such as efficient, centrally generated steam heat for neighborhoods. In Europe, housing codes and regulations often make it difficult for individuals to try new technologies. Overall, the world has not cut but has only slowed the growth of its energy consumption.

Both the growing demand for highly differentiated consumer goods and the effort to solve environmental problems create more jobs. Until the 1970s, it seemed that improved efficiency in agriculture and industry would lead either to massive unemployment or to a shorter workweek for all. Instead, in the last two decades of the twentieth century both the number of jobs and working hours increased. The new jobs were in the service and knowledge sectors of the economy. Productivity grew as corporations adopted lean production, flatter management structures, and computerized "just-in-time" inventory and delivery, but after 1975 hourly wages stagnated for many. In contrast, between c. 1860 and 1975 workers (particularly in Western Europe) gradually unionized and obtained a shorter workweek and higher wages. In recent decades, outsourcing of jobs has weakened unions, undermined job security, and pushed workers to accept longer hours. Even those with the best positions worked longer in 2005 than they did a generation ago. At the same time, many unions reinvented themselves to become partners with management, in some cases even sharing the same offices.

These choices in the workplace are inseparable from political and market choices. Older arguments about technology and the state often focused on the importance of a free press to democracy. But improved technologies of communication are not automatically shared or used by many people, nor are they necessarily used to investigate the government. Rupert Murdoch has assembled a media empire that includes newspapers, radio, television, and publishing. A single corporation owns the weekly magazine *Time,* the Warner studios, the Cable News Network, America Online, and much else.

Some recent films present dire visions of hegemonic corporate control by means of computers. *The Net* depicts a firm that sells security software to large firms and government offices.[10] With access to their databases, it manipulates information for its own advantage. It can affect the stock market, the Pentagon, or any institution that has become its client. It can change medical records, rewrite police files, invade bank accounts, or otherwise falsify data. The corporation has nearly completed its clandestine capture of the government when a young programmer begins to uncover its activities. In response, the corporation, using its access to bank records, police files, and other personal information, systematically destroys the programmer's identity. The corporation rewrites her curriculum vitae from that of a successful working woman into that of a prostitute and a drug addict wanted by the police. Meanwhile, another woman assumes her identity and her job. The programmer has the computer savvy to fight back and to expose the conspiracy. The average citizen identifies with her, but clearly lacks such skills.

If one danger is hegemony from the private sector, the equally daunting alternative is that the media can became a central apparatus of state control. In the satirical 1998 film *Wag the Dog*, a US president threatened by scandal in the midst of a re-election campaign restores his public image by hiring a "spin doctor."[11] He manipulates the media by inventing an imaginary overseas crisis that requires military intervention. He "documents" the invented crisis with "documentaries" fabricated in a studio. Such deception might seem far-fetched, but intelligence agencies have long used disinformation campaigns to confuse enemies, and some national leaders reiterate falsehoods on television with such sincerity that much of the public believes them.[12]

Possible misuse of the media is worrisome, but it is not the only issue. Even where the press is free of government controls or misuse and where political discussion is not fettered by monopoly, democratic nations have tended to let the marketplace make important technological decisions. The elected members of most legislatures have little technical or scientific expertise, and often the introduction of major new machines and processes is little discussed except by specialists. In the 1990s, relatively few people were involved in the decisions to commercialize the Internet, for example. Citizens were conceived primarily as passive consumers, not as voters.

Future developments in genetic engineering, in robotics, and in nanotechnology will force legislatures to make hard and potentially irreversible choices. One way to deal with government's technical illiteracy is to put engineers and scientists on congressional staffs. Another is to create advisory institutions, such as the Office of Technology of Assessment, which existed for a little more than two decades in the United States and inspired Europeans to create similar institutions. A third possible remedy for this problem is to bring representative groups of citizens together and crystallize their opinions through deliberative polling. With 3.3 million scientists and technicians working on research and development worldwide, some way for legislatures to make informed decisions about technology policy must be found.

Workers, voters, and consumers do not accept or reject technologies in isolation. Rather, all technologies are enmeshed in systems, and these may be closed or open. A closed system locks out competition and, by a variety of means, tries to keep consumers loyal. This can be done by creating a strong brand, by building in technological incompatibilities, or by creating synergies between the different elements of the system that make it more attractive than alternatives. Microsoft, for example, has used all of these techniques to dominate the computer software business. Technological systems are socially constructed; they can be open or closed, adaptable or rigid, democratically dispersed or restricted to elites.

In the future, will people have access to a variety of technical possibilities? Will they have the infrastructure needed to take advantage of potential choices? In the Netherlands, for example, many commuters have a real choice between the bicycle, mass transit, and the automobile. In contrast, the majority of Americans live in suburbs or rural areas with impoverished choices. The infrastructure is so distended that the automobile is the only practical option. The transportation choices available in most of the United States have contracted since 1905, when the country had millions of horses, a robust trolley system, bicycles, and some of the best passenger railways in the world. Today, most passenger trains have been abandoned and most interurban trolley lines have disappeared. In cities, the number of passengers on buses and trolleys dropped from 17.2 billion a year in 1950 to 7 billion a year in the early 1970s. Hundreds of smaller cities and towns abandoned public transportation.[13] Some suburbs do not even build sidewalks for walkers, and bicycle lanes are almost unheard of. The history of American transportation shows that a wealthy nation can make decisions that impoverish rather than enhance its choices.

Do people know enough to make choices? Does the citizen know how to ride a bicycle? Can the average person use the staggering array of options on a new home computer? Has the average consumer the skill needed to use the huge range of home improvement tools being sold? Cultures and organizations can seek to maximize skills, thereby enhancing resilience and flexibility. Henry Ford at first made a car that was simple enough for many people to repair. Today's automobiles are so complex that only highly trained mechanics can do so. Some early word processing software allowed users to add and delete fonts and utilities, so that owners could streamline and custom design software to fit their needs. In the 1980s it was possible to write a book on a computer with no hard disk, using software that required less than one megabyte. Two decades later, word processing programs are far less flexible but demand at least ten times as much memory. In contrast, Linux software is not sold as a technological "black box" with hidden code, but is an open-ended system that users can supplement with new code. Such open-source software empowers the consumer, but it takes more time to master. Consumers need to evaluate such tradeoffs. When manufacturers present new machines as "black boxes" which one cannot understand much less modify or repair, many people use them only until they need renovation or upgrading and then abandon them. When consumers give up on repairs, they can easily feel trapped, forced continually to upgrade their systems, as manufacturers determine the pace of change and the possible linkages between their machines. Fortunately, over the long term consumers generally seek variety, change, and independence.

The inability to understand or fix many modern machines is also linked to issues of safety. Even if each device functions perfectly by itself, ensembles of machines may have hidden incompatibilities that can cause malfunctions, accidents, or even disasters. Although each individual machine can be improved and made safer, the overarching system of machines may contain dangerous inconsistencies that manifest themselves only in extreme or unusual circumstances. Airplanes and high-speed trains usually function well and have a better safety record than automobiles. But a small malfunction can cause a devastating wreck at hundreds of miles per hour. The more powerful the system, the greater its destructive potential when it goes awry.

This general problem becomes even more acute in warfare, which places extreme demands on technologies and the people who use them. Modern armies have vastly improved their firepower and also their ability to protect soldiers. However, as it became more difficult to attack military targets, they attacked civilians instead. World War II was the turning point, and by its end bombers routinely destroyed entire cities. Dropping an atom bomb on Hiroshima was the tragic but "logical" result. By the end of the twentieth century, 90 percent of those who died in wars were civilians. Fatal mistakes became so common that military journalists invented euphemisms such as "friendly fire" and "collateral damage" to refer to killing one's own troops or unarmed civilians. An army tries to minimize such mistakes, but enemies will try to disrupt technological systems—for example, by jamming radios, decoying air strikes away from intended targets, or infecting computer software.

Extensive military training with computer simulations can transform actual battle into an almost unreal repetition of an arcade game. This disassociation began in the Renaissance, when artillery targets first were too far away to be clearly seen. For 400 years the distances have been increasing. Now a computerized system of representation mediates between the people who die on the ground and those who release bombs from miles in the air or fire a cruise missile at a target 100 miles away. Success or failure is abstract, represented by icons and movements on a screen. Such combatants are in a far different psychological state than those who engage in hand-to-hand battle. When simulated combat and real combat begin to look much the same, soldiers have entered a different experiential world. As a journalist observed, "the illusion is that we can win wars without killing people. Interactive video games that surpass even Gulf War television images threaten to revolutionize warfare further, by desensitizing tomorrow's policy makers completely to the consequences of fighting."[14]

There are equally important psychological implications to incorporating new technologies into everyday life. The experiences that seem natural to children today are radically unlike those of 200 years ago. A green lawn seems "normal" but it is artificially flattened, fertilized, and clipped, and could scarcely be found anywhere in 1805. The "normal" home in Western society invented since then has expanded to include indoor plumbing, central heating, hot running water, electric lighting, radio, refrigeration, television, and much more. The world that seems natural at our birth has been continually modified. One should be skeptical about claims that people can be easily or radically altered because they watch television, use the Internet, acquire a mobile phone, or purchase an intelligent machine. Nevertheless, the cumulative effect of continual innovation has encouraged people to see the world less as a shared dwelling than as a stockpile of raw materials. Technological peoples can unconsciously assume that the world exists for their convenience. The typical motorist assumes that gasoline at reasonable prices is "natural," and on the evening news it has become regrettable but "natural" to see a target from the vantage point of the nose cone of a missile, just as it has become "natural" to see bloody civilian casualties.

For those who surf the Internet for hours every day, using chat rooms and assuming different online personalities has also become "natural," moving them a vast distance from Martin Heidegger's position. He feared that people were losing touch with direct sensory experience of the natural world. In contrast, some people today *want* to leave their bodies behind and merge with the machine. Ray Kurzweil confidently asserted in *The Age of Spiritual Machines* that "there won't be mortality by the end of the twenty-first century [if] you take advantage of the twenty-first century's brain-porting technology."[15] For Kurzweil, future people will not be limited by their physical selves or by the hardware inherited from the past. Rather, he argues, eventually people will exist in their software, and they will migrate to new sites and expand their capabilities as computers become smaller and more powerful. These new sites might look like the bodies that emerged through biological evolution, but they will be rebuilt and improved using nanotechnologies.[16] For those who embrace this perspective, Heidegger's worries may seem mere nostalgia. Kurzweil's writings extravagantly illustrate how people increasingly treat the world as a mere storehouse of raw materials that can be used indiscriminately to satisfy their desires. Is all of nature, even the human body, merely a standing reserve awaiting exploitation and "improvement"?

Freeman Dyson predicts that our descendants will invent and develop radio telepathy so that they can experience collective memory and collective consciousness. He suggests they might break down the barriers between species and communicate with dolphins, whales, and chimpanzees. Conceivably, "those who have been part of an immortal group-mind may find it difficult to communicate with ordinary minds."[17] John Perry Barlow, in contrast, suggests that individual consciousness probably will survive: "Even though I believe that our advances in telecommunications are creating a great Mind that will combine all of our minds, I don't believe that individual human personalities will be subsumed into this vast organism."[18] Most people do not embrace either of these visions, and they are far less sanguine about computers. *New Yorker* cartoons suggest that many feel inferior to advanced machines. In one, a fortune teller warns a middle-aged businessman that he will never catch up with the new technology. In another, two robots stand looking at a museum display, which explains how they came to replace people. This history of evolution has five stages: a naked man, a typewriter, a mainframe computer of the 1970s, a personal computer, and (walking out of the frame to the right) a robot. A third cartoon shows a kitchen, where a man complains that he does not want to play chess with his microwave oven but only wants to warm up his lasagna.[19] Below the surface of these jokes lurks the fear that we are inadequate relative to the machine. We might become obsolescent.

Arthur C. Clarke argued that human invention ultimately can only lead to our evolutionary replacement by intelligent machines: "The tool we have invented is our successor. . . . The machine is going to take over."[20] For him, technology was not merely a series of ever more complex gadgets; it was the artificial extension and acceleration of evolution. Indeed, teams of researchers now actively seek not merely faster computers with more memory, but artificial intelligence. One researcher put it this way: "We're getting more and more alienated from these things [nature] that created us. The distance between us and what made us is growing very fast." However, he added, "from an evolutionary point of view, from a rational point of view . . . it doesn't matter whether the process [life or evolution] is carried on by carbon chemistry or by silicon or by robots. . . ."[21] Indeed, the status of intelligent machines is becoming a legal issue. Would an artificial intelligence deserve the same rights before the law that people have? Would an intelligent machine have the "right" to be eternally plugged in? Might ownership of life forms with artificial intelligence someday seem as heinous as slavery?

If machines can surpass humanity, will they take control? In some films, computer systems become malevolent. In Stanley Kubrick's film *2001,* a computer on board a space ship intentionally kills most of the crew. In the *Matrix* trilogy, a vast computer system pacifies most of humanity, keeping them asleep by linking them to a computer simulation that appears to be the urban world of the late twentieth century. In reality, it is several centuries later, after a terribly destructive world war. The surface of the earth is uninhabitable. The computer system "grows" people in pods in order to harvest their energy. They experience only a virtual reality that computers generate to pacify them, except for the few who reject the programming and escape into a reality far shabbier than the collective hallucination. Even the fact that some people reject the programming is a technological artifact, in that their rebellion is due to software errors at least as much as it expresses the human love of liberty. A different idea underlies the film *Lawnmower Man,* in which a scientist uses a combination of powerful drugs and computer programs to transform a stupid gardener into a man of superhuman intelligence. In the climactic scene, his powerfully enhanced mind merges with the research center's computer system and then becomes the core of a disembodied brain that bestrides the global computer network. Though many people will find either of these scenarios an unacceptable apocalypse, some contend that either merging with the machine or being replaced by it must be the logical result of history.

In contrast, the burden of my argument has been that there is no single, no logical, and no necessary end to the symbiosis between people and machines. For millennia, people have used tools to shape themselves and their cultures. We have developed technologies to increase our physical power, to perform all kinds of work, to protect ourselves, to produce surpluses, to enhance memory, and to extend perception. We have also excelled in finding new uses for inventions, and this has had many unexpected and not always welcome consequences. We are not necessarily evolving toward a single world culture, nor must we become subservient to (or extinct in favor of) intelligent machines. For millennia we have used technologies to create new possibilities. This is not an automatic process; it can lead either to greater differentiation or to increasing homogeneity. We need to consider the questions that technology raises because we have many possible futures, some far less attractive than others. We must "try to love the questions themselves like locked rooms and like books that are written in a very foreign tongue." As Rilke suggests, we may then "gradually, without noticing it, live along some distant day into the answers."[22] By refusing to let any ensemble of objects define our world as already given, we can continue to choose how technology matters.

Notes

Chapter 1

1. "Ancient hobbit-sized human species discovered," Associated Press, October 27, 2004.

2. I will not try to make this argument, but some scholars contend that the brain developed in interaction with tool use and therefore should be considered a human technology. See e.g. Beniger 1986, p. 9.

3. José Ortega y Gasset, "Man the Technician," in Ortega y Gasset 1941, p. 100.

4. Benjamin 1986, p. 93.

5. Whitney 1990, pp. 50–51.

6. Aristotle, *Nicomachean Ethics,* book 6, chapters 3 and 4.

7. *The Ethics of Aristotle* (Penguin Classics, 1953), p. 175.

8. Strauss 1959, p. 298.

9. Ibid.

10. Pavlovskis 1973, pp. 20 and 33, passim.

11. Cited on p. 52 of Whitney 1990.

12. Whitney 1990, pp. 139–140.

13. Ibid., pp. 143–145.

14. Sibley 1973, p. 264. See also Wallace 2003, pp. 11–21.

15. Cyril Stanley Smith, cited on p. 331 of Rhodes 1999.

16. MacKenzie 1998.

17. Smith, cited on p. 331 of Rhodes 1999.

18. Don Ihde, "The historical-ontological priority of technology over science," in Ihde 1983.

19. On the Wright Brothers, see Tobin 2003.

20. Bigelow 1840.

21. This statement is based on a survey of 100,000 nineteenth-century journal articles in Cornell University's electronic archive "Making of America" (available at http://moa.cit.cornell.edu/moa/index.html). Before 1840 there are only 34 uses of the term, all but three either in writings by Bigelow or in references to them. Two referred to curricula in German universities, and the last was an eccentric usage in a legal context that seems unrelated to machines.

22. Abraham Lincoln, "Second Lecture on Discoveries and Inventions," in *Collected Works of Abraham Lincoln,* volume 3, ed. R. Brasler (Rutgers University Press, 1953–55), pp. 357–358.

23. Oldenziel 2003.

24. Marx 1997.

25. Schatzberg, "Technik Comes to America."

26. Ibid.

27. Ibid.

28. Mumford 1934, p. 110.

29. Ibid., p. 100.

30. Bennett 1996.

31. Herlihy 1990, pp. 75–97.

32. Oldenziel 1999.

Chapter 2

1. On the rejection of technological determinism, see Winner 1977 and Smith and Marx 1994.

2. Discussed on pp. 78–79 and 188–189 of Basalla 1988.

3. Since 1985, more than two-thirds of the articles published in *Technology and Culture* have employed some form of a contextualist approach. For an overview, see Staudenmaier 1994.

4. Meyrowitz 1985, p. 309.

5. Negroponte 1995, p. 230.

6. Basalla 1988, p. 11. The classic study is Bulliet 1975.

7. Braudel 1973, p. 274.

8. Tenner 1996.

9. Ibid., pp. 220–223.

10. Ibid., pp. 207–208.

11. For a critical yet sympathetic summary and analysis of Marx, see pp. 461–482 of Sibley 1970.

12. Marx 1964, p. 64.

13. Marx 1997, p. 975.

14. See Schatzberg, "Technik Comes to America."

15. Cited on p. 51 of Feenberg 1999.

16. On Lenin's enthusiasm for electrification, see pp. 258–261 of Hughes 1989.

17. The Steinmetz citation is from p. 9 of Kline 1985.

18. Ogburn 1934, p. 124. For an insightful retrospective on Ogburn, see Volti 2004.

19. Ogburn 1934, p. 125.

20. Ogburn 1964, pp. 132–133.

21. Ibid, p. 133.

22. Toffler 1970. These ideas were developed further in Toffler 1980. For a more academic treatment of similar themes, see p. 309 of Meyrowitz 1985.

23. See Clark 1987.

24. Ellul 1970.

25. Cited on p. 61 of Winner 1977. In his little-read later work *The Ethics of Freedom* (1976), Ellul constructed a Christian argument that defied the technological system, building upon the work of Søren Kierkegaard and the existentialists.

26. Roszak 1969, pp. 7–8.

27. Ibid., chapters 4, 6, 7.

28. Foucault 1977, 1995.

29. Kurzweil 1980, p. 209.

30. Marx 1995, pp. 24–25.

31. Winner 1977.

32. Winner 1986, pp. 14–15.

Chapter 3

1. Utterback 1994, p. 193.

2. There are many examples of machines becoming much more popular after designers re-create them. See Meikle 1979.

3. CNN Evening News, European edition, October 12, 1998.

4. Barlow 2004, p. 177.

5. On how corporations exhibited themselves at fairs, see pp. 249–311 of Marchand 1998.

6. Wise 1976.

7. From http://web.jf.intel.com, October 20, 1998: "In 1965, Gordon Moore was preparing a speech and made a memorable observation. When he started to graph data about the growth in memory chip performance, he realized there was a striking trend. Each new chip contained roughly twice as much capacity as its predecessor, and each chip was released within 18-24 months of the previous chip. If this trend continued, he reasoned, computing power would rise exponentially over relatively brief periods of time. Moore's observation, now known as Moore's Law, described a trend that has continued and is still remarkably accurate. It is the basis for many planners' performance forecasts. In 26 years the number of transistors on a chip has increased more than 3,200 times, from 2,300 on the 4004 in 1971 to 7.5 million on the Pentium II processor."

8. This discussion began in the 1950s—see Raskin 1955. Kurt Vonnegut's *Player Piano* (1952) is a fictional exploration of these possibilities. For a general discussion, see pp. 4-5 of Schor 1991. For a pessimistic assessment, see Eric Hoffer, "Automation Is Here to Liberate Us," *New York Times Magazine,* October 24, 1965 (reprinted in Moore 1972).

9. Ehrlich 1968.

10. In 1902, when there were only a few thousand automobiles in the country, Americans took 4.8 billion trolley trips and made a billion transfers (Nye 1990, p. 96).

11. See Corn 1983. For an advertisement from the 1930s showing a "typical" family in a small plane over the Midwest, see p. 348 of Marchand 1998. On the perception of flight as sublime, see pp. 201-203 of Nye 1994.

12. Cited on p. 166 of Weart 1988.

13. Utterback 1994, p. 192.

14. However, overall energy use was still increasing at a rate of 50% per decade in the 1970s. For discussion, see chapter 8 of Nye 1998a.

15. Paulos 1995, p. 98. For a critique of Toffler and Naisbitt, see pp. 165-178 of Segal 1994.

16. Utterback 1994, p. 195.

17. Another example of the same sort of paralysis within a leading firm would be how Xerox failed to exploit its own development of the Windows program.

18. Lepartito 2003.

19. Blonheim 1994, p. 31; Israel 1992, p. 39.

20. Admittedly, the case of the telephone is complex, as Western Union hired Thomas Edison to create another version of the telephone, which he succeeded in doing. See Carlson 1994. For further development of these ideas, see Carlson 1998.

21. On the discovery itself, see pp. 660–674 of Rosenberg et al. 1995.

22. The source here is a formal document signed by Edison and his two principal assistants, Charles Batchelor and John Kruesi. See Rosenberg et al. 1995, p. 686.

23. I am indebted to Leo Marx for this anecdote, which stems from the mid 1970s, when he had recently taken up a post in MIT's program in Science, Technology, and Society. Subsequently, Langdon Winner confirmed it.

24. Another example would be cable television, which was around on a small and local scale for about 20 years without attracting much interest. I thank Eric Guthey for drawing my attention to this example.

25. Barlow 2004, p. 181.

26. See Bob Schwabach, "Home Users Lead the Way in Growth of Video Mail," Universal Press Syndicate story, published in *Minneapolis Star Tribune,* November 5, 1998.

27. The cost of computer memory has long been falling, but the cost of much software remains high. From the manufacturing point of view, software development remains expensive because programs keep increasing in complexity. This fact was recognized early—see Boehm 1973.

28. George Eastman figured out something similar in the 1890s: there was more money to be made selling film and processing than in selling cameras (Utterback 1994, pp. 175–176).

29. When the first transatlantic telephone service came on line, in the 1920s, a call cost $5 a minute—and $5 was a good day's pay. As with electricity, the basic cost of service declined as a result of technical improvements and economies of scale.

30. Photographs in the General Electric Photographic Archives at the Hall of History in Schenectady document two of these products. See negatives 506119 (toilet seat) and 281181 (sprinkler).

31. See Cusumano et al. 1992.

32. Utterback 1994, p. 28.

33. For a brief summary of how RCA refused to adopt FM and hindered its diffusion in the 1930s and the 1940s, see pp. 146–150 of Hughes 1989.

34. In Muncie, Indiana, district heating served more than 400 customers in the central business district during the 1920s. This business gradually declined and was abandoned in the 1960s as a result of interlinked economic and technological factors that encouraged the construction of power plants far from potential steam customers. See Frank Smikel, "Last Steam to Hiss through Downtown Lines Next Tuesday," *Muncie Star,* May 26, 1966.

35. See pp. 318–322 of Josephson 1959.

36. Czitrom 1982, pp. 68–72.

37. Another example from Muncie makes the same point. The interurban trolley cars sometimes made a point of getting a little ahead of schedule, so that they had time to stop on a bridge over a stream, drop an electrified wire into the water, and electrocute enough fish for dinner.

38. Hackett and Lotsenhiser 1985.

39. *General Electric Review* 1 (1903), November, p. 17.

40. Nye 1990, p. 60.

41. Ibid., pp. 49–50.

Chapter 4

1. Petroski 2003, p. 13.

2. For an elegant discussion, see Staudenmaier 1985.

3. Cited on pp. 57–58 of Staudenmaier 1985.

4. Wallace 2003, p. 3.

5. Susan B. Anthony, *New York World* interview, 1896, cited on p. 130 of Dodge 1996.

6. Hughes 1983, pp. 14–17.

7. Hughes 1969; Hughes 1994, p. 111.

8. Hughes 1989, p. 460.

9. Hughes 1994, p. 108.

10. Individual machines can achieve a shorter-term technological momentum, but only a few technologies achieve the lasting technological momentum of the railway network or the electrical system.

11. Hughes 1983, p. 140.

12. Ibid., p. 465.

13. Hughes 1994, p. 108.

14. Staudenmaier 1994, pp. 260–273.

15. See Kubler 1962.

16. Singer et al. 1951–1958.

17. Kidder 1995.

18. Israel 1998, pp. 410–421.

19. Hunter and Bryant 1991, pp. 61–68, 115–238.

20. Dukert 1980, p. 50.

21. Contextualists are the largest group of authors in Society for the History of Technology's journal *Technology and Culture*.

22. Scharff 1991, pp. 35–50.

23. Nye 1979, p. 23.

24. Ling 1990, p. 30.

25. Yergin 1992, pp. 208–209.

26. Lepartito 2003, p. 76.

27. Abbate 1999, pp. 43–81.

28. Ibid., pp. 83–144.

29. Furuland 1984, p. 38.

30. Renfrew 1984.

31. Luckin 1990, pp. 61, 68.

32. Nye 1990, p. 299. Electrification of farms was uneven, with 61% electrified in California in 1935 but less than 5% in most of the South. (Williams 1997, p. 223).

33. Powers 1992, p. 298.

34. McGaw 2003, p. 18.

Chapter 5

1. Dyson et al. 1994.

2. Cited on p. 196 of Winner 1997.

3. Huizinga 1972, pp. 234, 237.

4. On the Frankfurt School, see Jay 1973.

5. Quoted on p. 214 of Jay 1973.

6. Riesman 1950.

7. Grant 1969, p. 26.

8. Henry 1963, p. 15.

9. The often-reprinted speech can be found at www.narhist.ewu.edu.

10. Roszak 1969, p. 12.

11. This paragraph relies on pp. 234–238 of Jackson 1985.

12. Riesman 1958, pp. 375–402.

13. The information on the transformation of the original suburb assembled by Peter Bacon Hales can be seen at http://tigger.uic.edu.

14. "Behind the Region's Run-up in Prices," *New York Times* Sunday Late Edition, July 13, 2003.

15. Rae 1965, p. 61.

16. Ibid., pp. 95–99; Tedlow 1990, pp. 158–181.

17. Source: www.smartmoney.com.

18. This paragraph is based on an excellent paper presented by Kenneth Lepartito at the 1996 annual meeting of the Society for the History of Technology in London.

19. Blaszczyk 2000, p. 13.

20. Ibid., p. 93.

21. Ibid., p. 229.

22. Scranton 1997, p. 17.

23. Ibid., p. 99.

24. See e.g. Lederer 1961. For a Marxist version of this argument, see Ewen 1975.

25. Lyotard 1984, p. 4.

26. Castells 2001, p. 249.

27. Ibid., p. 250.

28. Ibid., p. 252.

29. Ibid., p. 261.

30. Miller and Slater 2001.

31. Ritzer 1993.

32. Barber 1996.

33. Fukuyama 2001. For a related argument, see Friedman 2000.

34. Robertson 1992. See also Appadurai 1996.

35. Kroes 1996, p. 164.

36. Durning 1992, pp. 69, 74, 68, 45.

37. Hobsbawm and Ranger 1983, p. 6.

38. Ibid., p. 4.

Chapter 6

1. David Humphreys, cited on p. 28 of Kasson 1976.

2. By "liberal thinkers" I mean those who embraced the idea of progress in the nineteenth and twentieth centuries, including utilitarians and advocates of free markets.

3. Marcuse 1970, p. 68.

4. Jackson 1984, p. 8.

5. See p. 114 of Nye 1998a.

6. Thurston 1881, pp. 7–8.

7. Thurston 1895, p. 310.

8. Sombart 1976, pp. 102–106.

9. Ibid., pp. 74, 87, 92.

10. Cited on p. 122 of Tichi 1987.

11. Tichi 1987, pp. 118–131.

12. Henry Ford, cited on p. 87 of Rhodes 1999.

13. Nye 1998b.

14. Greenhalgh 1988, pp. 52–81.

15. *Official Guide: Book of the Fair, 1933* (Chicago: A Century of Progress, 1933), p. 12.

16. Raskin 1955, p. 17.

17. Gibbs 1989, pp. 59–60.

18. White 1949, pp. 363–368.

19. Smith 1983.

20. *Los Angeles Herald-Examiner,* July 20, 1969.

21. Dyson et al. 1994.

22. Ibid., p. 3.

23. See e.g. Limerick 1988.

24. See pp. 12–13 of Long 2003.

25. The remaining examples in this paragraph are from pp. 75–85 of Gimpel 1988.

26. McNeill 2000, pp. 38–39.

27. Ibid., p. 48.

28. Berg 1980, p. 261.

29. Carlyle 1829, p. 266.

30. Ibid., p. 265.

31. Ibid.

32. Adams 1931, p. 380.

33. See chapter 11 of Nye 2003.

34. McNeill 2000, pp. 310–311.

35. Durning 1992, p. 82 and passim.

36. Sibley 1973, p. 267

37. Ibid., p. 261.

38. Defoe 1719, p. 127.

39. Ibid., p. 187.

40. In view of the inanity of many mobile phone conversations, it may be that people still seek improved means to unimproved ends.

41. Thoreau 1854, pp. 29, 42.

42. Blonheim 1994, pp. 32, 36.

43. Petroski 1989.

44. Gross 2000, pp. 181–196.

45. Elizabeth Hall Witherell, with Elizabeth Dubrulle, "The Life and Times of Henry D. Thoreau," at www.niulib.niu.edu.

46. Shi 1985.

47. Wachtel 1983.

48. Schumacher 1973.

49. Lovins 1976.

50. See e.g. World Watch Institute 2004.

51. Merchant 1989, p. 36.

52. White 1995, p. 112.

53. Cronon 1996, p. 455.

54. White 1996, p. 184.

55. Cohen 1995, pp. 368–369.

56. Ibid., p. 283.

Chapter 7

1. For a useful discussion of this point, see pp. 114–120 of Staudenmaier 1985.

2. Pacey 1999, pp. 18–20.

3. Galloway 2003, pp. 41–42.

4. On Robert Owen, see Morton 1962.

5. Smith 1980, pp. 106–107, 241–243, and passim.

6. See p. 44 of Cooper 1998.

7. The classic works are Thompson 1967 and Gutman 1977.

8. Hughes 1854, pp. 286–291.

9. Tocqueville 1840, volume 2, p. 171.

10. Taylor 1911.

11. Gramsci 1971, pp. 279–287, 303–313.

12. Montgomery 1976, p. 485.

13. Montgomery 1987.

14. Akin 1977.

15. On factory tourism, see pp. 127–131 of Nye 1994.

16. Slichter 1919; Brissenden and Frankel 1922.

17. Pursell 1994, pp. 115–116.

18. Milkman 1997, p. 148.

19. Ibid., pp. 7–8, 149–150.

20. Rifkin 1996, p. 59.

21. Ibid., p. xvii.

22. Pitkin 1932, p. 356.

23. Bix 2000, pp. 184–187.

24. Ibid., p. 225.

25. This view of the effects of technology's future consequences was a common theme in dystopian science fiction.

26. Cited in Noble 1984.

27. Sabel 1982, p. 57.

28. Ehrenreich 1989, pp. 206–207.

29. For an authoritative analysis of this process, see Shaiken 1986.

30. See Zuboff 1988.

31. Cited on p. 197 of Pursell 1994.

32. Nissen 2003, pp. 157–159.

33. Ibid., p. 159.

34. Zuboff 1988, p. 344.

35. Ibid., p. 348.

36. Ibid., p. 351.

37. Sabel 1982, p. 221.

38. Ibid., p. 223.

39. Drucker 1985.

40. See chapter 5 above.

41. Scranton 1977.

42. "2004 Ask a Working Woman Survey Report" (AFL-CIO, 2004), p. 4.

43. Long 2003, p. 58.

44. "Evidence from Census 2000 about Earnings by Detailed Occupation for Men and Women" (www.census.gov).

45. Cited on p. 482 of Light 1999.

46. "Evidence from Census 2000 about Earnings by Detailed Occupation for Men and Women" (www.census.gov), p. 9, figure 4.

47. Cowan 1983, pp. 26–31 and 208–219, passim.

48. Williams 2002, p. 201.

49. Humphries 1995, pp. 92–98.

50. Rifkin 1996.

51. Schor 1991, p. 40.

52. "Tema: Den Store Jobflugt," *Information,* April 1, 2004, p. 7.

53. Schlosser 2002, pp. 56, 60.

54. Leidner 1993, p. 120.

55. Ibid., p. 123.

56. Cited in ibid., p. 178.

57. "Grocery Workers Relieved, If Not Happy, at Strike's End," *New York Times,* February 28, 2004; "Wal-Mart Rejects 2000 Labor Audit," *USA Today,* January 14, 2004.

58. Exchange rates and price differences complicate minimum wage comparisons, but the US-EU contrast is marked.

59. For a global analysis of the problems American workers face, see pp. 208–224 of Reich 1992.

60. *Statistical Abstract of the United States, 2003* (Government Printing Office, 2003).

61. Schor 1998, p. 20.

62. Reich 1992, p. 208.

63. Hayes 1989, pp. 138–142.

Chapter 8

1. Wade Roush, "Genetic Savings and Clone," *Technology Review,* March 2005, p. 17.

2. "Dear FDA: Get Well Soon," *Technology Review,* March 2005, p. 45.

3. "The Laws," book 7 in *The Collected Dialogues of Plato* (Princeton University Press, 1963), pp. 1369–1370.

4. See pp. 97–101 of Pursell 1995.

5. Everett 1850, pp. 52–53.

6. Hughes 2004, pp. 27–29.

7. Mowry and Rosenberg 1998, pp. 21–22.

8. "OECD in Figures, 2004," available at www.oecd.org.

9. Electronic versions of every report from the Office of Technology Assessment are available at www.wws.princeton.edu.

10. Margolis and Guston 2003, pp. 68–69.

11. Hill 2003, p. 108.

12. Sibley 1973, p. 278.

13. Tony Blair, "Science Matters," speech to Royal Society, May 23, 2002.

14. Ibid.

15. *Official Guide, New York World's Fair, 1964/1965* (Time Inc., 1964), pp. 220–222; Let's Go to the Fair and Futurama (pamphlet, copy in Hagley Library, Hagley Museum, Delaware).

16. See www.nirs.org.

17. Davies 1975, pp. 20–23.

18. Goddard 1994, pp. 183–194.

19. Nye 1998a, pp. 217–234; Greenberger 1983.

20. Weinberg 1981. For a discussion of Weinberg, see Tatum 1996.

21. See "Drinking and Driving in Europe: Eurocare Report to the European Union" at www.eurocare.org.

22. Klein 2001.

23. Perlin 1999, p. 187.

24. Mander and Grossman 1997.

25. Berry 2000, p. 19. See also Berry 1981.

26. Berry 2000, p. 20.

27. Sclove 1995.

28. Goldman 1992, p. 149.

29. Sibley 1973, p. 278.

30. Hughes 2004, p. 41.

31. Fukuyama 2002, pp. 72–102.

32. Leon Kass, at http://bioethics.gov.

33. Fishman 1989; Winner 2001.

34. John Milton, "Areopagitica: Speech for the Liberty of Unlicensed Printing" (1664).

35. Pool 1983, p. 5.

36. Smith 1979, p. 105.

37. The rest of this paragraph is based on pp. 56–57 and 158–160 of Lewis 2002.

38. Pool 1983, p. 5.

39. On media concentration, see *The Nation,* January 7–14, 2002.

40. Samuel Huntington argued during the Watergate Hearings that a mass-media society could lead to too much democracy. He worried that a press and citizenry might become hyperactive, continually tying up the government in hearings and investigations. "The effective operation of a democratic political system requires some measure of apathy and non-involvement on the part of some individuals and groups." (Schudsen 1995, p. 220)

41. Gitlin 2001, p. 165.

42. Putnam 2000.

43. Goldman 1992, p. 150.

44. Habermas 1989, p. 247.

45. Rheingold 1993, 119–120.

46. Ibid., p. 113–114.

47. Marcuse 1969, p. 97.

48. Morgan and Peha, "Analysis, Governance, and the Need for Better Institutional Arrangements," in Morgan and Peha 2003, p. 9.

49. Cited in ibid., p. 9.

50. Stine 1994.

51. Vig and Paschen 2000.

52. James Fishkin and Robert C. Luskin, "Experimenting with a Democratic Ideal: Deliberative Polling and Public Opinion," presented to European University Institute, Florence, 2004. Available at http://cdd.stanford.edu.

53. Lehr et al. 2003.

54. Rosenberg et al. 1995, p. 686.

Chapter 9

1. Mukerji 2003, p. 659.

2. Schivelbusch 1986, p. 131.

3. *Louisiana Chronicle,* reprinted in *Boston Advertiser,* October 4, 1843; cited on p. 288 of Hunter 1949.

4. McCullough 1968, p. 193.

5. Chiles 2001, p. 303.

6. Turner 1978.

7. Lovins 1982, p. 197.

8. Ibid., p. 189.

9. Tenner 1997, p. 103.

10. Steinberg 2000, p. xvi.

11. Ibid., p. 47.

12. Ravetz 2003, p. 813.

13. Discussed on pp. 211–212 of Manion and Evan 2002.

14. Pauli Andersen, "Gift fra gen-planter skal undersøges," *Berlingske Tidende,* March 27, 2004.

15. Ravetz 2003, p. 814.

16. Manion and Evan 2002, p. 211.

17. Ibid., pp. 208–209.

18. See pp. 289–291 of Jensen 1996.

19. "Rates of Homicide, Suicide, and Firearm-Related Death among Children—26 Industrialized Countries," available at www.cdc.gov.

20. Miller et al. 2002.

21. Tenner 1997, pp. 45–47.

22. Bateman 1999, p. 10.

23. Perry 2004, p. 236.

24. Headrick 1981, pp. 43–54 (China) and 60–75 (Africa).

25. Ibid., pp. 117–118.

26. See www.loc.gov.

27. Browning 2002, p. 168. Preparing for World War II, the German military sponsored development of its famous panzer tank divisions and the rockets that would terrorize London. That conflict also saw the first use of radar, the deployment of 270,000 self-propelled artillery pieces and tanks, and the first atomic bombs.

28. Adas 1989, p. 365.

29. Headrick 1981, p. 124.

30. Mumford 1934, p. 310.

31. Antal 1999, p. 162.

32. Kennedy 1999, p. 603.

33. Ibid., pp. 604–610.

34. Harwell and Hutchinson 1985, p. 446.

35. China lost 2 million soldiers and more than 10 million civilians; Germany 3.5 million soldiers and 2 million civilians; Russia 10 million soldiers and 7 million civilians. The lower American casualty rate vindicated the policy of developing new weapons systems. Furthermore, in the war the United States had not lost a single factory, but rather expanded its industrial capacity, while producing 299,000 airplanes, 2.6 million machine guns, and 41 billion rounds of ammunition. See Kennedy 1999, p. 655; Jacobsen 2000, p. 79.

36. Winkler 1993, pp. 21–22; Hughes 1989, pp. 420–421.

37. Harwell and Hutchinson 1985, p. 447 (Dresden), p. 453 (Hiroshima).

38. This watch is on exhibit at the museum in Hiroshima.

39. On the coordination of large projects, see Hughes 2000.

40. Franklin 1988, chapter 1.

41. Orville Wright, cited on p. 66 of Rhodes 1999.

42. Cited on p. 33 of Franklin 1988.

43. Tesla 1900, p. 184.

44. For a summary of recent American military technology, see Perry 2004.

45. Kennan 1984, p. 172.

46. Jacobsen 2000, p. 78. For more detail, see Levy and Sidel 1997. Also see p. 2204 of Spiegel and Salama 2000.

47. Mumford 1946.

48. Kennan 1984, pp. 171–172.

49. Hecht 1998.

50. Cited on p. 281 of Weart 1988.

51. Atomic Energy Commission 1957.

52. See Kaldor 1982.

53. See Bruce Kennedy, "Nuclear Close Calls," at www.cnn.com. See also Hilgartner et al., pp. 214-215.

54. See p. 177 of Winkler 1993.

55. Behar 2002.

56. Hembree 2004.

57. Barnaby 1986, p. 155.

58. Antal 1999, p. 160.

59. Herby 2003, pp. 9–12.

60. Ibid., p. 12.

61. On the telegraph, see p. 10 of Czitrom 1982. On radio, see p. 306 of Douglas 1987. On television, see pp. 20–23 of Barnouw 1990. See also Winner 2004.

62. Russell 1961, p. 727.

63. Berkowitz 1987, p. 125.

64. McPhee 1974, p. 124.

Chapter 10

1. Emily Dickinson, poem A660, originally part of a letter to Samuel Bowles. Available at http://mith2.umd.edu.

2. Matt Richtel, "Wired to an Addiction: Multitaskers Get—and Need—the Rush," *International Herald Tribune,* July 7, 2003.

3. Ibid.

4. Debra Galant, "Driven to Distraction," *New York Times,* late edition, July 20, 2003.

5. Debbie Howlett, "Americans Driving to Distraction as Multitasking on Road Rises," *USA Today,* March 5, 2004.

6. Ibid.

7. Turkle 2004.

8. "The King Is Dead," *The Nation,* December 14, 1927

9. "Strut Miss Lizzie," *The Nation,* December 14, 1927.

10. Emerson 1950, p. 328.

11. Miller 1965, p. 306.

12. Ortega y Gasset 1941, p. 95.

13. Ibid., p. 153.

14. See Morton 2000; Braun 2002.

15. For more on technology and music, see Pinch and Bijsterveld 2003.

16. Heidegger 1977, p. 48.

17. Fitch 1975, p. 145.

18. Marvin 1988, pp. 93–95.

19. Peters 1999, p. 205.

20. Gitlin 2001, p. 206.

21. Jenkins 1994.

22. Goudie 1994, p. 184.

23. Winner 1986, p. 12.

24. Forster 1997.

25. J. C. R. Licklider, "The Computer as a Communication Device," cited in Rhodes 1999.

26. See www.atariarchives.org.

27. Frisch 1987, p. 178.

28. Haworth 2000, p. 59.

29. Mitchem 2000, p. 144.

30. Ibid., p. 146.

31. Strong 1992.

32. Connor 1997, p. 162.

33. Stone 1995.

34. See e.g. Deleuze and Guattari 1987.

35. Turkle 1997, p. 1103. Also see Turkle 1984, 1995.

36. Turkle 2004.

37. Hacking 2002; Showalter 1997.

38. Cited on p. 161 of Showalter 1997.

39. Showalter 1997, p. 161.

40. Dick 1968.

41. Sobchak, pp. 155–156.

42. Gibson 1986.

43. Haraway 2004, p. 9. See also Haraway 1996.

44. Gibson 1984, p. 67.

45. Gibson 1996.

Chapter 11

1. H. G. Wells, *When the Sleeper Wakes,* reprinted in *Three Prophetic Science Fiction Novels of H. G. Wells* (Dover, 1960); Aldous Huxley, *Brave New World* (Perennial reprint, 1998); George Orwell, *1984* (Signet reprint, 1990). Increases in computer speed and memory make it far easier today to establish Orwell's universal surveillance, while the rapid advances in biology seem to make Huxley's genetic engineering possible.

2. There are other, more complex ways to divide up historians of technology, but this is not an essay in methodology. Staudenmaier 1985 remains a good starting point.

3. Davis 1990.

4. Fiske 1993, p. 52.

5. Rolston 1998.

6. United Nations Environmental Programme 1997, p. 218.

7. Source: www.nrdc.org.

8. Ibid.

9. *Our Common Future* 1987, pp. 8, 89. For discussion, see Duchin et al. 1990. See also Duchin and Lange 1994.

10. *The Net,* directed by Irwin Winkler (Columbia Pictures, 1995).

11. *Wag the Dog,* directed by Barry Levinson (New Line, 1998).

12. Lyndon Johnson gained additional war powers after he convinced Congress and the American people that their navy had been attacked in the Bay of Tonkin, but little evidence suggests such an attack took place. In 2002, George W. Bush convinced many of the American people, without any hard evidence, that Iraq possessed weapons of mass destruction and that its leaders were closely linked to international terror networks.

13. Hilton 1985, pp. 45–47.

14. Trevor Corson, "Try Out Smart Bombs of Tomorrow in Living Rooms Today," *Christian Science Monitor,* July 6, 1998.

15. Kurzweil 1999, p. 128.

16. Ibid., pp. 134–136 and 140, passim.

17. Dyson 1997, pp. 157–158.

18. Barlow 2004, p. 184.

19. These and many other cartoons dealing with technology and the future can be found at www.cartoonbank.com.

20. Clarke 1999, p. 194

21. Cited on p. 199 of Helmreich 1998.

22. Rilke 1954, p. 35.

Bibliography

Abbate, Janet. 1999. *Inventing the Internet*. MIT Press.

Adams, Henry. 1931. *The Education of Henry Adams*. Modern Library.

Adas, Michael. 1989. *Machines as the Measure of Man*. Cornell University Press.

Akin, William E. 1977. *Technocracy and the American Dream: The Technocracy Movement, 1900–1941*. University of California Press.

Antal, John. 1999. "The End of Maneuver." In *Digital War: A View from the Front Line*, ed. R. Bateman. Presidio.

Appadurai, Arjun. 1996. *Modernity at Large*. University of Minnesota Press.

Atomic Energy Commission. 1957. Theoretical Possibilities and Consequences of Major Accidents in Large Nuclear Power Plants.

Barber, Benjamin. 1996. *Jihad vs. McWorld*. Ballantine.

Barlow, John Perry. 2004. "The Future of Prediction." In *Technological Visions: The Hopes and Fears That Shape New Technologies*, ed. M. Sturken et al. Temple University Press.

Barnaby, Frank. 1986. *The Automated Battlefield*. Sidgwick and Jackson.

Barnouw, Erik. 1990. *Tube of Plenty*. Oxford University Press.

Basalla, George. 1988. *The Evolution of Technology*. Cambridge University Press.

Bateman, Robert. 1999. "Pandora's Box." In *Digital War: A View from the Front Lines,* ed. R. Bateman. Presidio.

Behar, Michael. 2002. "The New Mobile Infantry." *Wired* 10, no. 5.

Beniger, James R. 1986. *The Control Revolution: Technological and Economic Origins of the Information Society.* Harvard University Press.

Benjamin, Walter. 1986. *Reflections: Essays, Aphorisms, Autobiographical Writings,* ed. P. Demetz. Shocken.

Bennett, Judith M. 1996. *Ale, Beer and the Brewster in England: Women's Work in a Changing World, 1300–1600.* Oxford University Press.

Berg, Maxine. 1980. *The Machinery Question and the Making of Political Economy, 1815–1848.* Cambridge University Press.

Berkowitz, Bruce. 1987. *Calculated Risks.* Simon & Schuster.

Berry, Wendell. 1981. *The Gift of Good Land.* Farrar, Straus and Giroux.

Berry, Wendell. 2000. "Back to the Land." In *The Best American Science and Nature Writing 2000,* ed. D. Quammen. Houghton Mifflin.

Bigelow, Jacob. 1840. *The Useful Arts.* Thomas Webb.

Bix, Amy Sue. 2000. *Inventing Ourselves Out of Jobs? America's Debate over Technological Unemployment, 1929–1981.* Johns Hopkins University Press.

Blaszczyk, Regina Lee. 2000. *Imagining Consumers: Design and Innovation from Wedgewood to Corning.* Johns Hopkins University Press.

Blonheim, Menahem. 1994. *News over the Wires: The Telegraph and the Flow of Public Information in America, 1844–1897.* Harvard University Press.

Boehm, Barry. 1973. "Software and Its Impact: A Quantitative Assessment." *Datamation* 19, no. 5: 48–59.

Braudel, Fernand. 1973. *Capitalism and Material Life: 1400–1800.* Harper Torchbooks.

Braun, Hans-Joachim. 2002. *"I Sing the Body Electric": Music and Technology in the Twentieth Century.* Johns Hopkins University Press.

Brissenden, Paul F., and Eli Frankel. 1922. *Labor Turnover in Industry: A Statistical Analysis.* Macmillan.

Browning, Peter. 2002. *The Changing Nature of Warfare.* Cambridge University Press.

Bulliet, Richard W. 1975. *The Camel and the Wheel.* Harvard University Press.

Carlson, W. Bernard. 1994. "Entrepreneurship in the Early Development of the Telephone: How Did William Orton and Gardiner Hubbard Conceptualize This New Technology?" *Business and Economic History* 23, winter: 161–192.

Carlson, W. Bernard. 1998. "Taking On the World: Bell, Edison, and the Diffusion of the Telephone in the 1870s." Presented at "Prometheus Wired" conference, Munich

Carlyle, Thomas. 1829. "Signs of the Times." *Edinburgh Review* 49, June: 438–459.

Castells, Manuel. 2001. *The Internet Galaxy: Reflections on the Internet, Business, and Society.* Oxford University Press.

Chiles, James. 2001. *Inviting Disaster: Lessons from the Edge of Technology.* HarperCollins.

Clark, Gregory. 1987. "Why Isn't the Whole World Developed? Lessons from the Cotton Mills." *Journal of Economic History* 47, no. 1: 141–173.

Clarke, Arthur C. 1999. *Profiles of the Future.* HarperCollins.

Cohen, Joel E. 1995. *How Many People Can the Earth Support?* Norton.

Connor, Steven. 1997. "Feel the Noise: Excess, Affect and the Acoustic." In *Emotion in Postmodernism,* ed. G. Hoffmann and Alfred Hornung. Universitätsverlag C. Winter.

Cooper, Gail. 1998. *Air-Conditioning America: Engineers and the Controlled Environment, 1900–1960.* Johns Hopkins University Press.

Corn, Joseph. 1983. *America's Romance with Aviation, 1900–1950.* Oxford University Press.

Cowan, Ruth Schwartz. 1983. *More Work for Mother.* Basic Books.

Cronon, William, ed. 1996. *Uncommon Ground.* Norton.

Cusumano, Michael A., Yiorgos Mylonadis, and Richard S. Rosenbloom. 1992. "Strategic Maneuvering and Mass-Market Dynamics: The Triumph of VHS over Betamax." *Business History Review* 66, spring: 51–94.

Czitrom, Daniel J. 1982. *Media and the American Mind.* University of North Carolina.

Davies, Richard O. 1975. *The Age of Asphalt: The Automobile, the Freeway, and the Condition of Metropolitan America.* Lippincott.

Davis, Mike. 1990. *City of Quartz.* Verso.

Defoe, Daniel. 1719. *Robinson Crusoe.* Reprint: Barnes and Noble, 2003.

Deleuze, Gilles, and Felix Guattari. 1987. *A Thousand Plateaus: Capitalism and Schizophrenia.* University of Minnesota Press.

Dick, Philip K. 1968. *Do Androids Dream of Electric Sheep?* Ballantine.

Dodge, Pryor. 1996. *The Bicycle.* Flammarion.

Douglas, Susan. 1987. *Inventing American Broadcasting,* 1899–1922. Johns Hopkins University Press.

Drucker, Peter F. 1985. *Innovation and Entrepreneurship.* HarperCollins.

Duchin, Faye, and F. G. Lange. 1994. *Ecological Economics.* Oxford University Press.

Duchin, Faye, F. G. Lange, and T. Johnson. 1990. *Strategies for Environmentally Sound Development: An Input-Output Analysis.* Third Progress Report to the United Nations.

Dukert, Joseph M. 1980. *A Short Energy History of the United States.* Edison Electric Institute.

Durning, Alan. 1992. *How Much Is Enough? The Consumer Society and the Future of the Earth.* Norton.

Dyson, Esther, George Gilder, George Keyworth, and Alvin Toffler. 1994. *Cyberspace and the American Dream: A Magna Carta for the Knowledge Age,* Release 1.2, August 22.

Dyson, Freeman. 1997. *Imagined Worlds.* Harvard University Press.

Ehrenreich, Barbara. 1989. *Fear of Falling: The Inner Life of the Middle Class.* Harper.

Ehrlich, Paul. 1968. *The Population Bomb.* Ballantine.

Ellul, Jacques. 1970. *The Technological Society.* Knopf.

Ellul, Jacques. 1976. *The Ethics of Freedom.* Eerdmans.

Emerson, Ralph Waldo. 1950. "The Poet." In *Selected Writings of Emerson,* ed. B. Atkinson. Modern Library.

Everett, Edward. 1850. "Address on Fourth of July at Lowell." In *Orations and Speeches of Edward Everett,* volume 2. Charles C. Little and James Brown.

Ewen, Stuart. 1975. *Captains of Consciousness: Advertising and the Social Roots of the Consumer Culture.* McGraw-Hill.

Feenberg, Andrew. 1999. *Critical Theory of Technology.* Routledge.

Fishman, Rachel. 1989. "Patenting Human Beings: Do Sub-Human Creatures Deserve Constitutional Protection?" *American Journal of Law and Medicine* 15, no. 4: 461–482.

Fiske, John. 1993. *Power Plays, Power Works.* Verso.

Fitch, James Marsden. 1975. *American Building 2: The Environmental Forces That Shaped It.* Schocken.

Forster, E. M. 1997. *The Machine Stops and Other Stories.* Andre Deutsch.

Foucault, Michel. 1977. *Discipline and Punish: The Birth of the Prison.* Viking.

Foucault, Michel. 1995. *The Birth of the Clinic: An Archeology of Medical Perception.* Vintage Books.

Franklin, H. Bruce. 1988. *War Stars: The Superweapon and the American Imagination.* Oxford University Press.

Friedman, Thomas. 2000. *The Lexus and the Olive Tree*. Anchor Books.

Frisch, Max. 1987. *Homo Faber*. Harcourt Brace.

Fukuyama, Francis. 2001. "Social Capital, Civil Society, and Development." *Third World Quarterly* 22: 7–20.

Fukuyama, Francis. 2002. *Our Posthuman Future: Consequences of the Biotechnology Revolution*. Farrar, Straus, and Giroux.

Furuland, Lars. 1984. "Ljus over landet: Elektrifieringen och litteraturen." *Daedalus*. Tekniska Museets Arsbok.

Galloway, Colin. 2003. *One Long Winter Count*. University of Nebraska Press.

Gibbs, Nancy. 1989. "How America Has Run Out of Time." *Time,* April 24: 59–60.

Gibson, William. 1984. *Neuromancer*. HarperCollins.

Gibson, William. 1986. *Burning Chrome*. HarperCollins.

Gibson, William. 1996. *Idoru*. Putnam.

Gimpel, Jean. 1988. *The Medieval Machine: The Industrial Revolution of the Middle Ages,* second edition. Wildwood House.

Gitlin, Todd. 2001. *Media Unlimited: How the Torrent of Images and Sounds Overwhelms Our Lives*. Holt.

Goddard, Stephen B. 1994. *Getting There: The Epic Struggle between Road and Rail in the American Century*. University of Chicago Press.

Goldman, Steven L. 1992. "No Innovation without Representation: Technological Action in a Democratic Society." In *New Worlds, New Technologies, New Issues,* ed. S. Cutcliffe et al. Lehigh University Press.

Goudie, Andrew. 1994. *The Human Impact on the Natural Environment,* fourth edition. MIT Press.

Gramsci, Antonio. 1971. *Selections from the Prison Notebooks*. International Publishers.

Grant, George. 1969. *Technology and Empire*. House of Anasi.

Greenberger, Martin. 1983. *Caught Unawares: The Energy Debate in Retrospect.* Ballinger.

Greenhalgh, Paul. 1988. *Ephemeral Vistas: The Expositions Universelles, Great Exhibitions and World's Fairs, 1851–1939.* Manchester University Press.

Gross, Robert. 2000. "'That Terrible Thoreau': Concord and Its Hermit." In *A Historical Guide to Henry David Thoreau,* ed. W. Cain. Oxford University Press.

Gutman, Herbert. 1977. *Work, Culture, and Society in Industrializing America.* Random House.

Habermas, Jürgen. 1989. *The Structural Transformation of the Public Sphere: An Inquiry into a Category of Bourgeois Society.* MIT Press.

Hackett, Bruce, and Loren Lotsenhiser. 1985. "The Unity of Self and Object." *Western Folklore* 44: 317–324.

Hacking, Ian. 2002. *Rewriting the Soul: Multiple Personality and the Sciences of Memory.* Diane.

Haraway, Donna. 1996. *Modest Witness @ Second Millennium.* Routledge.

Haraway, Donna. 2004. "A Manifesto for Cyborgs." In *The Harraway Reader.* Routledge.

Harwell, Mark A., and Thomas C. Hutchinson. 1985. *Environmental Consequences of Nuclear War,* volume 2. Wiley.

Haworth, Lawrence. 2000. "Focal Things and Focal Practices." In *Technology and the Good Life,* ed. E. Higgs et al. University of Chicago Press.

Hayes, Dennis. 1989. *Behind the Silicon Curtain: The Seductions of Work in a Lonely Era.* Free Association Books.

Headrick, David R. 1981. *The Tools of Empire: Technology and European Imperialism in the Nineteenth Century.* Oxford University Press.

Hecht, Gabrielle. 1998. *The Radiance of France: Nuclear Power and National Identity after World War II.* MIT Press.

Heidegger, Martin. 1977. "The Turning." In *The Question Concerning Technology and Other Essays.* Harper and Row.

Helmreich, Stefan. 1998. *Silicon Second Nature.* University of California Press.

Hembree, Amy. 2004. "The Things They'll Carry." *Wired* 12, no. 2.

Henry, Jules. 1963. *Culture Against Man.* Random House.

Herby, Peter. 2003. "Grim Future." *The World Today,* May: 9–12.

Herlihy, David. 1990. *Opera Muliebria: Women and Work in Medieval Europe.* Temple University Press.

Hilgartner, Stephen, Richard C. Bell, and Rory O'Connor. 1983. *Nukespeak: The Selling of Nuclear Technology in America.* Penguin.

Hill, Christopher P. 2003. "Expanded Analytical Capability in CRS, GAO or CBO." In *Science and Technology Advice for Congress,* ed. M. Morgan and J. Peha. Resources for the Future.

Hilton, George W. 1985. "The Rise and Fall of Monopolized Transit." In *Urban Transit,* ed. C. Lave. Ballinger.

Hobsbawn, Eric, and Terence Ranger, eds. 1983. *The Invention of Tradition.* Cambridge University Press.

Hughes, Henry. 1854. *A Treatise on Sociology, Theoretical and Practical.* Lippincott.

Hughes, James. 2004. *Citizen Cyborg: Why Democratic Societies Must Respond to the Redesigned Human of the Future.* Westview.

Hughes, Thomas P. 1969. "Technological Momentum: Hydrogenation in Germany, 1900–1933." *Past and Present,* August: 106–132.

Hughes, Thomas P. 1983. *Networks of Power: Electrification in Western Society, 1880–1930.* Johns Hopkins University Press.

Hughes, Thomas P. 1989. *American Genesis: A Century of Invention and Technological Enthusiasm.* Viking Penguin.

Hughes, Thomas P. 1994. "Technological Momentum." In *Does Technology Drive History?* ed. M. Smith and L. Marx. MIT Press.

Hughes, Thomas P. 2000. *Rescuing Prometheus: Four Monumental Projects That Changed the World*. Vintage.

Huizinga, Johan. 1972. *Life and Thought in America: A Dutch Historian's Vision, From Afar and Near*. Harper Torchbooks. Original publication: *Mensch en menigte in America* (H. D. Tjeenk Willink en Zoon, 1918).

Humphries, Jane. 1995. "Women and Paid Work." In *Women's History, Britain, 1850–1945,* ed. J. Purvis. University College London.

Hunter, Louis C. 1949. *Steamboats on Western Rivers*. Harvard University Press.

Hunter, Louis C., and Lynwood Bryant. 1991. *A History of Industrial Power in the United States, 1780–1930,* volume 3: *The Transmission of Power*. MIT Press.

Ihde, Don. 1983. *Existential Technics*. SUNY Press.

Israel, Paul. 1992. *From Machine Shop to Industrial Laboratory: Telegraphy and the Changing Context of American Invention, 1830–1920*. Johns Hopkins University Press.

Israel, Paul. 1998. *Edison: A Life of Invention*. Wiley.

Jackson, John Brinckerhoff. 1984. *Discovering the Vernacular Landscape*. Yale University Press.

Jackson, Kenneth T. 1985. *Crabgrass Frontier: The Suburbanization of the United States*. Oxford University Press.

Jacobsen, John Hurt. 2000. *Technical Fouls: Democratic Dilemmas and Technological Change*. Westview.

Jay, Martin. 1973. *The Dialectical Imagination*. Little, Brown.

Jenkins, Virginia Scott. 1994. *The Lawn: A History of an American Obsession*. Smithsonian Institution.

Jensen, Claus. 1996. *No Downlink: A Dramatic Narrative about the Challenger Accident and Our Time*. Farrar, Strauss, and Giroux.

Josephson, Matthew. 1959. *Edison: A Biography*. McGraw-Hill.

Kaldor, Mary. 1982. *The Baroque Arsenal.* London: Andre Deutsch.

Kasson, John F. 1976. *Civilizing the Machine: Technology and Republican Values in America, 1776–1900.* Reprint: Hill & Wang, 1999.

Kennan, George F. 1984. "American Diplomacy and the Military." In Kennan, *American Diplomacy,* expanded edition. University of Chicago Press.

Kennedy, David. 1999. *Freedom from Fear.* Oxford University Press.

Kidder, Tracey. 1995. *The Soul of a New Machine.* Avon.

Klein, Naomi. 2001. *No Logo.* Flamingo.

Kline, Ronald. 1985. "Electricity and Socialism: The Career of Charles P. Steinmetz." Presented at History of Technology Colloquium, University of Delaware.

Kroes, Rob. 1996. *If You've Seen One, You've Seen the Mall.* University of Illinois Press.

Kubler, George. 1962. *The Shape of Time: Remarks on the History of Things.* Yale University Press.

Kurzweil, Edith. 1980. *The Age of Structuralism: Levi-Strauss to Foucault.* Columbia University Press.

Kurzweil, Ray. 1999. *The Age of Spiritual Machines.* Penguin.

Lederer, William J. 1961. *A Nation of Sheep.* Fawcett.

Lehr, R. L., W. Guild, D. L. Thomas, and B. G. Swezey. 2003. Listening to Customers: How Deliberative Polling Helped Build 1,000 MW of New Renewable Energy Projects in Texas. National Renewable Energy Laboratory, Golden, Colorado.

Leidner, Robin. 1993. *Fast Food, Fast Talk: Service Work and the Routinization of Everyday Life.* University of California Press.

Lepartito, Kenneth. 2003. "Picturephone and the Information Age: The Social Meaning of Failure." *Technology and Culture* 44, no. 1: 50–81.

Levy, B., and V. Sidel. 1997. *War and Public Health.* Oxford University Press.

Lewis, Bernard. 2002. *What Went Wrong? Western Impact and Middle Eastern Response*. Weidenfield and Nicholson.

Light, Jennifer. 1999. "When Computers Were Women." *Technology and Culture* 40, no. 3: 455–483.

Limerick, Patricia. 1988. *The Legacy of Conquest*. Norton.

Ling, Peter. 1990. *America and the Automobile: Technology, Reform and Social Change, 1893–1923*. Manchester University Press.

Long, Pamela O. 2003. *Technology and Society in the Medieval Centuries: Byzantium, Islam, and the West, 500–1300*. American Historical Association.

Lovins, Amory. 1976. "Energy Strategy: The Road Not Taken." *Foreign Affairs* 55, October: 65–83.

Lovins, Amory. 1982. *Brittle Power: Energy Strategy for National Security*. Brick House.

Luckin, Bill. 1990. *Questions of Power: Electricity and Environment in Inter-War Britain*. Manchester University Press.

Lyotard, Francois. 1984. *The Postmodern Condition: A Report on Knowledge*. University of Minnesota Press.

MacKenzie, Donald. 1998. *Knowing Machines*. MIT Press.

Mander, Jerry, and Edward Grossman. 1997. *The Case against the Global Economy*. Sierra Club.

Manion, M., and W. N. Evan. 2002. "Technological Catastrophes: Their Causes and Prevention." *Technology in Society* 24: 207–224.

Marchand, Roland. 1998. *Creating the Corporate Soul: The Rise of Public Relations and Corporate Imagery in American Big Business*. University of California Press.

Marcuse, Herbert. 1969. "Repressive Tolerance." In *A Critique of Pure Tolerance*, ed. R. Wolff et al. Beacon.

Marcuse, Herbert. 1970. *Five Lectures*. Beacon.

Margolis, Robert M., and David H. Guston. 2003. "Origins, Accomplishments, and Demise of the OTA." In *Science and Technology Advice for Congres,* ed. M. Morgan and J. Peha. Resources for the Future.

Marvin, Carolyn. 1988. *When Old Technologies Were New.* Oxford University Press.

Marx, Karl. 1964. *Selected Writings in Sociology and Social Philosophy,* ed. T. Bottomore and M. Rubel. McGraw-Hill.

Marx, Leo. 1995. "The Idea of Technology and Postmodern Pessimism." In *Technology, Pessimism, and Postmodernism,* ed. Y. Ezrahi et al. University of Massachusetts Press.

Marx, Leo. 1997. "Technology: The Emergence of a Hazardous Concept." *Social Research* 64, no. 3: 965–988.

McCullough, David. 1968. *The Johnstown Flood.* Simon & Schuster.

McGaw, Judith. 2003. "Why Feminine Technologies Matter." In *Gender and Technology,* ed. N. Lerman, R. Oldenziel, and A. Mohun. Johns Hopkins University Press.

McGrew, W. C. 1993. *Chimpanzee Material Culture: Implications for Human Evolution.* Cambridge University Press.

McNeill, J. R. 2000. *Something New under the Sun: An Environmental History of the Twentieth Century World.* Norton.

McPhee, John. 1974. *The Curve of Binding Energy: A Journey into the Awesome and Alarming World of Theodore B. Taylor.* Farrar, Straus and Giroux.

Meikle, Jeffrey. 1979. *Twentieth Century Limited: Industrial Design in America, 1925–1939.* Temple University Press.

Merchant, Carolyn. 1989. *Ecological Revolutions: Nature, Gender and Science in New England.* University of North Carolina.

Meyrowitz, Joshua. 1985. *No Sense of Place: The Impact of Electronic Media on Social Behavior.* Oxford University Press.

Milkman, Ruth. 1997. *Farewell to the Factory: Auto Workers in the Late Twentieth Century.* University of California Press.

Miller, Daniel, and Don Slater. 2001. *The Internet: An Ethnographic Approach*. Berg.

Miller, M., D. Azrael, and D. Hemenway. 2002. "Firearm Availability and Unintentional Firearm Deaths, Suicide, and Homicide among 5–14 Year Olds." *Journal of Trauma* 52: 267–275.

Miller, Perry. 1965. *The Life of the Mind in America*. Harcourt, Brace and World.

Mitchem, Carl. 2000. "Of Character and Technology" In *Technology and the Good Life*, ed. E. Higgs et al. University of Chicago Press.

Montgomery, David. 1976. "Worker's Control of Machine Production in the Nineteenth Century." *Labor History* 17: 485–509.

Montgomery, David. 1987. *The Fall of the House of Labor: The Workplace, the State, and American Labor Activism, 1865–1925*. Cambridge University Press.

Moore, Wilbert E. 1972. *Technology and Social Change*. Quadrangle.

Morgan, M. Granger, and Jon M. Peha. 2003. *Science and Technology Advice for Congress*. Resources for the Future.

Morton, A. L. 1962. *The Life and Ideas of Robert Owen*. International Publishers.

Morton, David. 2000. *Off the Record: The Technology and Culture of Sound Recording in America*. Rutgers University Press.

Mowry, David C., and Nathan Rosenberg. 1998. *Paths of Innovation*. Cambridge University Press.

Mukerji, Chandra. 2003. "Intelligent Uses of Engineering and the Legitimacy of State Power." *Technology and Culture* 44, no. 4: 655–676.

Mumford, Lewis. 1934. *Technics and Civilization*. Harcourt, Brace.

Mumford, Lewis. 1946. "Gentlemen, You Are Mad!" *Saturday Review of Literature*, March 2: 5–6.

Negroponte, Nicholas. 1995. *Being Digital*. Vintage.

Nissen, Bruce. 2003. "What Are Scholars Telling the U.S. Labor Movement to Do?" *Labor History* 44, no. 2: 157–165.

Noble, David F. 1984. *Forces of Production.* Knopf.

Nye, David E. 1979. *Henry Ford: Ignorant Idealist.* Academic Press.

Nye, David E. 1990. *Electrifying America: Social Meanings of a New Technology.* MIT Press.

Nye, David E. 1994. *American Technological Sublime.* MIT Press.

Nye, David E. 1998a. *Consuming Power: A Social History of American Energies.* MIT Press.

Nye, David E. 1998b. "Electrifying Expositions." In Nye, *Narratives and Spaces.* Columbia University Press.

Nye, David E. 2003. *America as Second Creation: Technology and Narratives of New Beginnings.* MIT Press.

Ogburn, W. F. 1934. "The Influence of Invention and Discovery." In Ogburn, *Recent Social Trends in the United States.* McGraw-Hill.

Ogburn, W. F. 1964. *On Culture and Social Change: Selected Papers.* University of Chicago Press.

Oldenziel, Ruth. 1999. *Making Technology Masculine: Men, Women and Modern Machines in America.* University of Amsterdam Press.

Oldenziel, Ruth. 2003. "The Genealogy of Technology: Race, Gender, and Class." Presented at annual meeting of Society for the History of Technology, Atlanta.

Ortega y Gasset, José. 1941. *Toward a Philosophy of History.* Norton.

Our Common Future. 1987. World Conference on Environment and Development.

Pacey, Arnold. 1999. *Meaning in Technology.* MIT Press.

Paulos, John Allen. 1995. *A Mathematician Reads the Newspaper.* Basic Books.

Pavlovskis, Zoya. 1973. *Man in an Artificial Landscape: The Marvels of Civilization in Imperial Roman Literature.* Leiden: Mnemosyne, Biblioteca Batava.

Perlin, John. 1999. *From Space to Earth: The Story of Solar Electricity.* Harvard University Press.

Perry, William J. 2004. "Military Technology: An Historical Perspective." *Technology in Society* 26: 235–243.

Peters, John Durham. 1999. *Speaking into the Air: A History of the Idea of Communication.* University of Chicago Press.

Petroski, Henry. 1989. "H. D. Thoreau, Engineer." *American Heritage of Invention and Technology* 5, no. 2: 8–16.

Petroski, Henry. 2003. *Small Things Considered: Why There Is No Perfect Design.* Vintage.

Pinch, Trevor J., and Karin Bijsterveld. 2003. "'Should One Applaud?' Breaches and Boundaries in the Reception of New Technology in Music." *Technology and Culture* 44, no. 3: 536–559.

Pitkin, Walter B. 1932. *The Consumer: His Nature and Changing Habits.* McGraw-Hill.

Pool, Ithiel de Sola. 1983. *Technologies of Freedom.* Harvard University Press.

Powers, Richard. 1992. *Three Farmers on Their Way to a Dance.* Harper Perennial.

Pursell, Carroll. 1994. *White Heat: People and Technology.* University of California Press.

Pursell, Carroll. 1995. *The Machine in America.* Johns Hopkins University Press.

Putnam, Robert D. 2000. *Bowling Alone.* Simon and Schuster.

Rae, John B. 1965. *The American Automobile.* University of Chicago Press.

Raskin, A. H. 1955. "Pattern for Tomorrow's Industry?" *New York Times Magazine,* December 18, p. 17.

Ravetz, Jerry. 2003. "A paradoxical future for safety in the global knowledge economy." *Futures* 35: 811–826.

Reich, Robert B. 1992. *The Work of Nations.* Vintage.

Renfrew, Christie. 1984. *Electricity, Industry and Class in South Africa.* SUNY Press.

Rheingold, Howard. 1993. *The Virtual Community: Homesteading on the Electronic Frontier.* Addison-Wesley.

Rhodes, Richard, ed. 1999. *Visions of Technology.* Simon and Schuster.

Riesman, David. 1950. *The Lonely Crowd.* Yale University Press reprint, 2001.

Riesman, David. 1958. "The Suburban Sadness." In *The Suburban Community,* ed. W. Dobriner. Putnam.

Rifkin, Jeremy. 1996. *The End of Work: The Decline of the Global Labor Force and the Dawn of the Post-Market Era.* Putnam.

Rilke, Rainer Maria. 1954. *Letters to a Young Poet.* 1954.

Ritzer, George. 1993. *The McDonaldization of Society.* Pine Forge.

Robertson, Roland. 1992. *Globalization.* Sage.

Rolston, Holmes, III. 1998. "Technology versus Nature: What Is Natural?" *Ends and Means* 2, no. 2.

Rosenberg, Robert A., Paul B. Israel, Keith A. Nier, and Melodie Andrews. 1995. *Menlo Park: The Early Years, April 1876–December 1877, The Papers of Thomas A. Edison,* volume 3. Johns Hopkins University Press.

Roszak, Theodore. 1969. *The Making of a Counter Culture.* Doubleday.

Russell, Bertrand. 1961. "Open Letter to Eisenhower and Khrushchev." In Russell, *Basic Writings.* Allen and Unwin.

Sabel, Charles. 1982. *Work and Politics: The Division of Labor in Industry.* Cambridge University Press.

Scharff, Virginia. 1991. *Taking the Wheel: Women and the Coming of the Motor Age.* Free Press.

Schatzberg, Eric. 2003. "Technik Comes to America: The American Concept of Technology before World War II." Paper presented at annual meeting of Society for the History of Technology, Atlanta. Forthcoming in *Technology and Culture.*

Schivelbusch, Wolfgang. 1986. *The Railway Journey: The Industrialization of Time and Space in the Nineteenth Century.* Berg.

Schlosser, Eric. 2002. *Fast Food Nation: The Dark Side of the All-American Meal.* Perennial.

Schor, Juliet B. 1991. *The Overworked American: The Unexpected Decline of Leisure.* Basic Books.

Schor, Juliet B. 1998. *The Overspent American: Upscaling, Downshifting, and the New Consumer.* Basic Books.

Schudsen, Michael. 1995. *The Power of News.* Cambridge University Press.

Schumacher, E. F. 1973. *Small Is Beautiful: Economics as if People Mattered.* Harper and Row.

Sclove, Richard E. 1995. *Democracy and Technology.* Guilford.

Scranton, Philip. 1997. *Endless Novelty: Specialty Production and American Industrialization, 1865–1925.* Princeton University Press.

Segal, Howard P. 1994. *Future Imperfect: The Mixed Blessings of Technology in America.* University of Massachusetts Press.

Shaiken, Harley. 1986. *Work Transformed: Automation and Labor in the Computer Age.* Lexington Books.

Shi, David. 1985. *The Simple Life.* Oxford University Press.

Showalter, Elaine. 1997. *Hystories: Hysterical Epidemics and Modern Media.* Columbia University Press.

Sibley, Mulford Q. 1970. *Political Ideas and Ideologies: A History of Political Thought.* Harper & Row.

Sibley, Mulford Q. 1973. "Utopian Thought and Technology." *American Journal of Political Science* 37, no. 2: 255–281.

Singer, Charles, E. J. Holmyard, A. R. Hall, Trevor I. Williams, et al. 1951–1958. *A History of Technology.* Oxford University Press.

Slichter, Sumner L. 1919. *The Turnover of Factory Labor.* Appleton.

Smith, Anthony. 1979. *The Newspaper, An International History.* Thames and Hudson.

Smith, Merritt Roe. 1980. *Harper's Ferry Armory and the New Technology.* Cornell University Press.

Smith, Merritt Roe and Leo Marx, eds. 1994. *Does Technology Drive History? The Dilemma of Technological Determinism.* MIT Press.

Smith, Michael. 1983. "Selling the Moon: The US. Manned Space Program and the Triumph of Commodity Scientism." In *The Culture of Consumption,* ed. T. Lears and R. Fox. Pantheon.

Sobchak, Vivian. 2004. "Science Fiction Film and the Technological Imagination." In *Technological Visions: The Hopes and Fears That Shape New Technologies,* ed M. Sturken et al. Temple University Press.

Sombart, Werner. 1976. *Why Is There No Socialism in the United States?* M. E. Sharpe.

Spiegel, Paul B., and Peter Salama. 2000. "War and Mortality in Kosovo, 1998–99: An Epidemiological Testimony." *Lancet* 355, June: 2204–2209.

Staudenmaier, John M. 1985. *Technology's Storytellers: Renewing the Human Fabric.* MIT Press.

Staudenmaier, John M. 1994. "Rationality versus Contingency in the History of Technology." In *Does Technology Drive History? The Dilemma of Technological Determinism,* ed. M. Smith and L. Marx. MIT Press.

Steinberg, Ted. 2000. *Acts of God: The Unnatural History of Natural Disaster in America.* Oxford University Press.

Stine, Jeffery K. 1994. *Twenty Years of Science in the Public Interest.* American Association for the Advancement of Science.

Stone, Allucquère Rosanne. 1995. *The War of Desire and Technology at the Close of the Mechanical Age.* MIT Press.

Strauss, Leo. 1959. *Thoughts on Machiavelli*. Free Press.

Strong, David. 1992. "The Technological Subversion of Environmental Ethics." *Research in Philosophy and Technology: Technology and the Environment* 12: 33–66.

Sturken, Marita, Douglas Thomas, and Sandra J. Ball-Rokeach, eds. 2004. *Technological Visions: The Hopes and Fears That Shape New Technologies*. Temple University Press.

Tatum, J. S. 1996. "Technology and Liberty: Enriching the Conversation." *Technology in Society* 18, no. 1: 41–59.

Taylor, Frederick Winslow. 1911. *The Principles of Scientific Management*. Harper.

Tedlow, Richard S. 1990. *New and Improved*. Basic Books.

Tenner, Edward. 1997. *Why Things Bite Back: Technology and the Revenge of Unintended Consequences*. Vintage.

Tesla, Nicola. 1900. "The Problem of Increasing Human Energy with Special Reference to the Harnessing of the Sun's Energy." *Century Magazine,* May: 175–211.

Thompson, E. P. 1967. "Time, Work Discipline, and Industrial Capitalism." *Past and Present* 38, no. 1: 56–97.

Thoreau, Henry David. 1854. *Walden*. Reprint: Holt, Rinehart and Winston, 1948.

Thurston, Robert H. 1881. "Our Progress in Mechanical Engineering: The President's Annual Address." *Transactions of the American Society of Mechanical Engineers* (copy in Cornell University Library).

Thurston, Robert H. 1895. "The Trend of National Progress." *North American Review* 161, September.

Tichi, Cecelia. 1987. *Shifting Gears: Technology, Literature, Culture in Modernist America*. University of North Carolina Press.

Tobin, James. 2003. *To Conquer the Air: The Wright Brothers and the Great Race for Flight*. Free Press.

Tocqueville, Alexis de. 1840. *Democracy in America,* volume 2. Reprint: Vintage, 1945.

Toffler, Alvin. 1970. *Future Shock.* Random House.

Toffler, Alvin. 1980. *The Third Wave.* Bantam Books.

Turkle, Sherry. 1984. *The Second Self.* Simon and Schuster.

Turkle, Sherry. 1995. *Life on the Screen: Identity in the Age of the Internet.* Simon and Schuster.

Turkle, Sherry. 1997. "Computational Technologies and Images of the Self." *Social Research* 64, no. 3: 1093–1112.

Turkle, Sherry. 2004. "How Computers Change the Way We Think." *Chronicle of Higher Education* 50, no. 21: B26.

Turner, B. A. 1978. *Man-Made Disasters.* Wykeham Science Press.

United Nations Environmental Programme. 1997. *Global Environmental Outlook.* Distributed by Oxford University Press.

Utterback, James M. 1994. *Mastering the Dynamics of Innovation.* Harvard Business School.

Vig, Norman J., and Herbert Paschen. 2000. *Parliaments and Technology: The Development of Technology Assessment in Europe.* SUNY Press.

Volti, Rudi. 2004. "William F. Ogburn, *Social Change with Respect to Culture and Original Nature." Technology and Culture* 45, no. 2: 396–405.

Vonnegut, Kurt. 1952. *Player Piano.* Avon.

Wachtel, Paul. 1983. *The Poverty of Affluence.* Free Press.

Wallace, Anthony F. C. 2003. *The Social Context of Invention.* University of Nebraska Press.

Weart, Spencer R. 1988. *Nuclear Fear: A History.* Harvard University Press.

Weinberg, Alvin M. 1981. "Can Technology Replace Social Engineering?" In *Technology and Man's Future,* ed. A. Teich, third edition. St. Martin's Press.

White, Leslie A. 1949. *The Science of Culture*. Grove.

White, Richard. 1995. *The Organic Machine: The Remaking of the Columbia River*. Hill and Wang.

White, Richard. 1996. "Are You an Environmentalist or Do You Work for a Living? Work and Nature." In *Uncommon Ground,* ed. W. Cronon. Norton.

Whitney, Elspeth. 1990. *Paradise Restored: The Mechanical Arts from Antiquity through the Thirteenth Century*. American Philosophical Society.

Williams, James C. 1997. *Energy and the Making of Modern California*. University of Akron.

Williams, Rosalind. 2002. *Retooling: A Historian Confronts Technological Change*. MIT Press.

Winkler, Alan. 1993. *Life under a Cloud*. Oxford University Press.

Winner, Langdon. 1977. *Autonomous Technology: Technics-out-of-Control as a Theme in Political Thought*. MIT Press.

Winner, Langdon. 1986. *The Whale and the Reactor*. University of Chicago Press.

Winner, Langdon. 2001. "Are Humans Obsolete?" *Hedgehog Review* 3, no 3.

Winner, Langdon. 2004. "Sow's Ears from Silk Purses: The Strange Alchemy of Technological Visionaries." In *Technological Visions: The Hopes and Fears That Shape New Technologies,* ed. M. Sturken et al. Temple University Press.

Wise, George. 1976. Technological Prediction, 1890–1940. Ph.D. dissertation, Boston University.

World Watch Institute. 2004. *State of the World 2004*.

Yergin, Daniel. 1992. *The Prize: The Epic Quest for Oil, Money, and Power*. Simon and Schuster.

Zuboff, Shoshana. 1988. *In the Age of the Smart Machine: The Future of Work and Power*. Basic Books.

Index